Climate Change

A MULTIDISCIPLINARY APPROACH

WILLIAM JAMES BURROUGHS

CAMBRIDGE
UNIVERSITY PRESS

PUBLISHED BY THE PRESS SYNDICATE OF THE UNIVERSITY OF CAMBRIDGE
The Pitt Building, Trumpington Street, Cambridge, United Kingdom

CAMBRIDGE UNIVERSITY PRESS
The Edinburgh Building, Cambridge, CB2 2RU, UK
40 West 20th Street, New York, NY 10011–4211, USA
477 Williamstown Road, Port Melbourne, VIC 3207, Australia
Ruiz de Alarcón 13, 28014 Madrid, Spain
Dock House, The Waterfront, Cape Town 8001, South Africa

First published 2001
Reprinted 2002

Printed in the United States of America

Typeface Melior 9.75/13 and Eurostile   *System* 3B2   [KWP]

A catalog record for this book is available from the British Library

Library of Congress Cataloging-in-Publication Data
Burroughs, William James
Climate change : a multidisciplinary approach / William James Burroughs.
p.   cm.
ISBN 0-521-56125-6 – ISBN 0-521-56771-8
1. Climate changes. I. Title.
QC981.8.C5 B86 2001   00-033750
551.6–dc21

ISBN 0-521-56125-6 hardback
ISBN 0-521-56771-8 paperback

CONTENTS

PREFACE

We are all inclined to take the climate for granted. We are basically acclimatised to the seasonal cycle and most of the variations that occur from day to day and week to week. It is all too easy to forget just how much of this comfort depends on the fact that our buildings, food and energy supplies, health and transport systems and leisure activities are carefully designed to meet the challenges of the local climate. But, when extreme events occur, it becomes acutely apparent how vulnerable much of the infrastructure of society is to climatic fluctuations. Droughts, floods, heatwaves and windstorms (including hurricanes and tornadoes) can all have disastrous consequences.

This sense of independence is often sometimes compounded by our travels. If we jet around the world staying in air-conditioned hotels or indulging in well-equipped holiday facilities, again we can fall into the trap of assuming that we are relatively independent of the weather. If, however, we are stripped of our protective shell and forced to confront the extremes of the tropics, deserts, high mountains and polar wastes, we can become painfully aware of just how forbidding the challenges of many climatic zones are. What is more, for many of the people living in the developing world, where extreme weather events have become a mounting threat to their way of life, these challenges are of vital concern. So, on an increasingly crowded planet it is hardly surprising that the subject of climate change has become a hot topic.

What is surprising is that this interest in climate change is a relatively recent phenomenon. There are two principal reasons for this development. First, it was not until the work of a few dedicated researchers such as Hubert Lamb in England and J. Murray Mitchell in the USA, in the 1950s and 1960s, that the reality of recent climate change became an accepted scientific concept. This type of scholarly work set standards in the examination of various records to establish reliable evidence of climate change. It was also aided by the appreciation that not only was there a lot of information in both instrumental records and documentary sources, but also in the many other examples of environmental change, including tree rings, ice cores, pollen records and ocean sediments. The use of many of these sources depended in part on the emergence of new technologies that improved the quality of measurements and, hence, enabled more reliable inferences to be drawn about the past.

The second reason for increased interest has been the growing realisation in the last few decades that recent climate change may well be at least partly the result of human activities. If this is indeed the case, it has major implications for economic and social development. This awareness has grown on the back of a greater understanding of the mechanisms driving natural climate fluctuations. It is a measure of these developlments that Hubert Lamb's major two-volume work *Climate: Present, Past and Future*, published in the 1970s, contains very little mention of the El Niño and the Southern Oscillation (ENSO): it was not until the record-breaking event in 1982/3 that the true significance of this natural variation in the climate became apparent. Now it is seen as central to many aspects of interannual climate variability around the world.

The other essential component of climate studies has been the explosive growth of computer technology. This has led to many advances, of which perhaps the most important is the development of computer models that are capable of simulating many features of the global climate. The power of these models is such that it is now possible to explore various aspects of climate change that may result from natural causes, or as a result of human activities. What is more, in recent years a growing confidence has emerged that we may now be capable of predicting just what future changes may occur if human activities continue unabated. These predictions paint a worrying picture of our vulnerability to future changes.

This combination of the growing importance of climate change in our lives and our increasing ability to both unravel past variations and predict future developments make this an ideal time to take stock of all aspects of the subject. This is best done by exploring the subject in the round. It requires us to examine a whole range of disciplines. It starts with what the science of meteorology tells us about the climate of the Earth, and leads into the evidence of past changes and their consequences. It needs to assess the reliability of various measurement sciences and how they tease out the subtle details of changes on every timescale – from the recent past to the depths of geological time – and how statistics are used to extract the maximum amount of useful information from records. Armed with these insights it is possible to study the implications of these changes in history, economics, agriculture and other social and political areas, and to consider how predictions of future changes may impact on society.

Add to all this the intricate detective work that weaves together all the clues as to how and why the climate changes to form a coherent picture, and you have the ingredients for an intriguing story. Moreover, when it comes to assessing the impact of changes, from mass extinctions (e.g. the Death of the Dinosaurs) to the part played by human activities in global warming, the plot thickens. So, only by looking at all aspects of climate change can we form a balanced judgement about the relevance to widely debated contemporary environmental, economic and social issues. My objective in this book is to help you, the reader, to come to this balanced judgement.

William J. Burroughs

ACKNOWLEDGEMENTS

As this book draws on lengthy personal involvement in climate matters, it is difficult to identify all the people who have helped me to form a view on the many facets of climate, how it has changed and its impact on all our lives. Among the meteorological community I would like to thank: Chris Folland, David Parker, Tony Slingo, John Mitchell, Bruce Callendar, David Griggs, Jack Hopkins and Alan Thorpe at the UK Meteorological Office; David Anderson, Tim Palmer, Tony Hollingsworth, Peter Jannsen, Anders Peerson and Austen Woods, at the European Center for Medium Range Weather Forecasting; Grant Bigg, Keith Briffa, Tom Holt, Mike Hulme and Phil Jones at the University of East Anglia; David Cotton at the University of Southampton; John Harries and Joanna Haigh at Imperial College; David Blackman at the Proudman Oceanographic Laboratory; Chris Mutlow at the Rutherford Appleton Laboratory; Niel Lonie at the University of Dundee; Dominic Reeve at Sir William Halcrow and Partners Ltd; Christopher Landsea at NOAA Hurricane Research Division, Miami; David Robinson at Rutgers University; Tom Karl at NOAA Climate Data Center; John Christy at the University of Alabama; and Jan Lindstrom at the University of Helsinki. Thanks are also due to Franz Fliri and Norman Lynagh for helpful discussions, and the provision of data and other material, which in one way or another was essential for completion of the book. In addition, I am grateful to Richard Alley, Roger Barry, Per Gloersen, Michael Hambrey, Michael Mann, Martin Parry, Julia Slingo, Jon Snow, and Kevin Trenberth for helpful advice on climate matters.

I also wish to acknowledge the importance of two vital sources of information and inspiration for producing this book. First, there is the groundbreaking work of the late Hubert Lamb, and in particular his major two-volume opus *Climate: Present, Past and Future*, which was published in the 1970s and provides an ideal guide of what any book on climate change should address. Secondly, there is the work of the Intergovernmental Panel on Climate Change (IPCC), which has, since its establishment in 1988, produced a series of publications that represent a massive resource and by far the most comprehensive assessment of climate issues available at present. It is no accident that this book relies so heavily on this source for so many of its illustrations. Finally, I am deeply indebted to my wife, who, as always, helped and supported me throughout the lengthy gestation of this book.

LIST OF BOXES

CHAPTER ONE

INTRODUCTION

> There is always an easy solution to every human problem – neat plausible and wrong.
>
> H. L. Menken

The climate has always been changing. On every timescale, since the Earth was first formed its surface conditions have fluctuated. Past changes are etched on the landscape, have influenced the evolution of all lifeforms, and are a subtext of our economic and social history. Current changes are a central part of the debate about the consequences of human activities on the global environment, while the future course of the climate could exert powerful constraints on economic development, especially in developing countries. So for many physical and social sciences the subject of climate change is an underlying factor which needs to be appreciated in understanding how these disciplines fit in to the wider picture. The aim of this book is to provide a balanced view to assist the reader to give the right weight to the impact of climate change on their chosen disciplines. This will involve assessing how the climate can vary on its own accord and then adding in the question of how human activities may lead to further change.

The first thing to get straight is that there is nothing simple about how the climate changes. While the central objective of the book is to make the essence of the subject accessible, it is no help to you, the reader, to have an oversimplified presentation of the issues. While many people try to reduce the issues to what they see as the essential features, either unwittingly or deliberately, they run the risk of mis-interpreting the evidence. So, from the outset it pays to appreciate that the behaviour of the Earth's climate is governed by a wide range of factors all of which are interlinked in an intricate web of physical processes. This means we must start by identifying which factors matter most and when they come into play. To do this, we have to define the meaning of climate change because various factors assume different significance depending on the timescales under consideration.

1.1   WEATHER AND CLIMATE

The first stage in establishing what we mean by climate variability and climate change is to discriminate between weather and climate. At the simplest level the weather is what is happening to the atmosphere at any given time, while climate is what would be expected to occur at any given time of the year based on statistics built up over many years. Although climatic statistics concentrate on averages, analysis has to include the incidence of extreme events which are part of the normal for any part of the world. Here, the emphasis will be principally on the average conditions, but it is in the nature of statistics that giving proper weight to rare extreme events is difficult to handle. Wherever possible the analysis will be in terms of well-established statistical series, which enables the change in the incidence of extremes to be handled appropriately. Where the changing frequency of extreme weather events exerts a major influence on the interpretation of changes in the climate, however, there may be some blurring of the distinction between weather and climate in considering specific events. The essential point is, in considering climatic issues, we are concerned about the statistics of the weather phenomena which provide evidence of longer term changes.

1.2   WHAT IS CLIMATE VARIABILITY AND CLIMATE CHANGE?

It follows from the definition of weather and climate, that changes in the climate constitute shifts in meteorological conditions lasting a few years or longer. These changes may involve a single parameter, such as temperature or rainfall, but usually accompany more general shifts in weather patterns which might result in a shift to, say, colder, wetter, cloudier and windier conditions. Because of the connection with global weather patterns these changes can result in compensating shifts in different parts of the world. More often they are, however, part of an overall warming or cooling of the global climate. But in terms of considering the implications of changes in the climate, it is the regional variations which provide the most interesting material, as long as they are properly set in the context of global change.

This leads into the question of defining the difference between climate variability and climate change. Given that we will be considering a continuum of variations across the timescale from a few years to a billion years, there is bound to be a degree of artificiality in this distinction. So it is important to spell out clearly how the two categories will be treated in this book. Figure 1.1(a) presents a typical set of meteorological observations; this example is a series of annual average temperatures, but it could equally well be rainfall or some other meteorological variable for which regular measurements have been made over the years. This series shows that over the period of the measurements the average value remains effectively constant (the series is said to be *stationary*) but fluctuates considerably from observation to observation. This fluctuation about the mean is a measure of *climate variability*. In Fig. 1.1(b), (c) and (d) the same example of climatic variability is combined with examples of *climate change*. The combination of variability and a uniform cooling trend is shown in Fig. 1.1(b), while in curve (c) the variability is combined with a periodic change in the underlying climate, and in curve (d) the variability is combined with a sudden drop in temperature, which represents, during the period of observation, a once and for all change in the climate.

The implication of the forms of change shown in Fig. 1.1 is that the level of variability remains constant while the climate changes. This need not be the case. Figure 1.2 presents the implications of variability changing as well. Curve (a) presents the combination of the amplitude of variability doubling over the period of observation, while the climate remains constant. Although this is not a likely scenario, the possibility of the variability increasing as, say, the climate cools (Fig. 1.2(b)) is much more likely. Similarly, the marked increase in variability following a sudden drop in temperature (Fig. 1.2(c)) is a possible consequence of climate change. The examples of climate variability and climate change presented in Figs. 1.1 and 1.2 will be explored in this book, and so we will return to the concepts in these diagrams from time to time.

Detecting fluctuations in the climate of the type described in Figs. 1.1 and 1.2 involves measuring a range of past variations of meteorological parameters around the world over a wide variety of timescales. This poses major challenges which will be a central theme in this book. For now, the important point is that when we talk about climate variability and climate change we are dealing with evidence which comes from a comprehensive range of sources and this varies greatly in quality. So, some past changes stand out with startling clarity, as in the case of the broad features of the last ice age, whereas others, such as the effect of sunspots on more recent changes, are shadowy and surrounded by controversy. It is this mixture of the well established and the unknown, which make the subject so hard to pin down and so fascinating.

1.3 CONNECTIONS, TIMESCALES AND UNCERTAINTIES

From the outset it is essential to realise that it does not help to oversimplify the workings of the climate. In truth it is an immensely complicated subject. Part of the excitement of discovering the implications of climate change is,

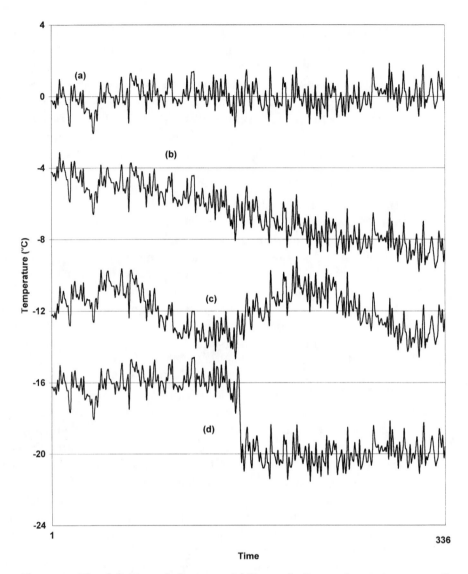

Figure 1.1 The definition of climate variability and climate change is most easily presented by considering a typical set of temperature observations which show **(a)** climate variability without any underlying change in the climate, **(b)** the combination of the same climate variability with a linear decline in temperature of 4 °C over the period of the observations, **(c)** the combination of climate variability with a periodic variation in temperature of 3 °C, and **(d)** the combination of climate variability with a sudden drop in temperature of 4 °C during the record, with the average temperature otherwise remaining constant before and after the shift. Each record is displaced by 4 °C to enable a comparison to be made more easily.

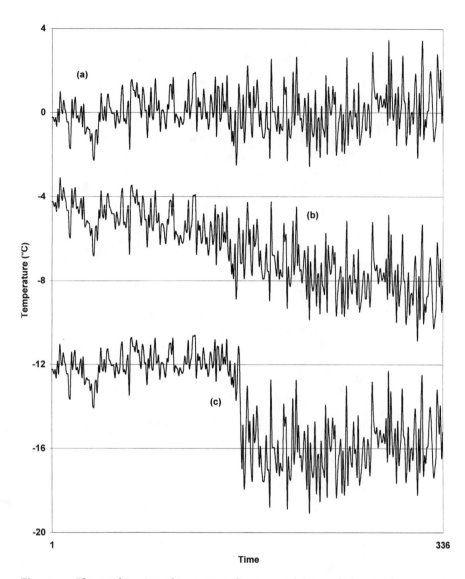

Figure 1.2 The combination of increasing climate variability and climate change can be presented by considering a set of temperature observations similar to those in Fig. 1.1 which show **(a)** climate variability doubling over the period of the record without any underlying change in the climate, **(b)** the combination of the same increasing climate variability with a linear decline in temperature of 4 °C over the period of the observations, **(c)** the combination of one level of climate variability before a sudden drop in temperature of 4 °C, which then doubles after the drop while with the average temperature remains constant before and after the shift. Each record is displaced by 4 °C to enable a comparison to be made more easily, and the example of periodic climate change in Fig. 1.1 (curve c) is not reproduced here as the nature of changing variability in such circumstances is likely to be more complicated.

BOX 1.1   THE NATURE OF FEEDBACK

Understanding and quantifying various *feedback processes* is a central challenge of explaining and predicting climate change. These processes arise because, when one climatic variable changes, it alters another in a way that influences the initial variable, which triggered the change. If this circular response leads to a reinforcement of the impact of the original stimulus then the whole system may move dramatically in a given direction. This runaway response is known as *positive feedback* and is best illustrated by the high pitch whistle that a microphone system can produce if it picks up some of its own signal. In the case of the climate, an example of this behaviour might be the effect of a warming leading to a reduction in snow cover in winter. This, in turn, could lead to more sunlight being absorbed at the surface and yet more warming, and so on.

The reverse situation is when the circular response tends to damp down the impact of the initial stimulus and produces a steady state. This is known as *negative feedback*. An example of this type of climatic response is where a warming leads to more water vapour in the atmosphere, which produces more clouds. These reflect more sunlight into space thereby reducing the amount of heating of the surface and so tend to cancel out the initial warming. Depending on the complexity of the system involved, these types of chain reaction can lead to a variety of responses to any perturbations, including sudden switches between different states, and reversals where a positive stimulus can produce a negative reaction by the system as a whole. Throughout this book the appreciation of feedback mechanisms will play an essential part in understanding how the climate responds to change.

however, the fact that it involves so many different processes linked together in an intricate web. In particular it requires us to understand the nature of *feedback processes* (see Box 1.1). The fact that a perturbation in one part of the system may produce effects elsewhere, which bear no simple relation to the original stimulus provides new insights into how the world around us functions. It does require, however, a disciplined approach to the physical processes at work, otherwise the analysis is liable to be partial and misleading.

The challenge of discovering which processes matter most involves not only knowing how a given perturbation may disturb the climate but also of how different timescales affect the analysis of change. From the imperceptibly slow movement of the continents to the day-to-day flickering of the Sun, every aspect of the forces driving the Earth's climate are varying. Deciding which of these changes matters most, and when, requires us to evaluate how rapidly they occur and how they are linked to one another. So while continental drift (*plate tectonics*) only comes into play when interpreting geological records over millions of years its more immediate consequences (e.g. volcanism) can have a sudden and dramatic impact on the variability of the climate. Similarly, fluctuations in the output of the Sun may occur on every timescale. When it comes to looking into the future, longer term variations are bound to be the subject of speculation and, as such, are of questionable value. But, in the case of predictions a few days to several decades ahead, the potential benefits of accurate forecasts are immense, and so this is one of the hottest topics in climate change.

This differing perspective, depending on whether the analysis of the past is relevant to forecasting the future, will influence how climatic change is presented in this book. To use an example, the oil industry exploits the growing information about past climates on geological timescales to assist in the search for hydrocarbon deposits. For instance, current research includes computer models of past ocean circulation, which can be used to identify regions of high marine productivity in the geologic past, where large amounts of organic matter may have been laid down. But changing knowledge about the climate on these timescales will have no influence on the industry's view on how the climate in the future will affect its plans.

By way of contrast, as production extends farther offshore, the oil industry needs to know whether changes in, say, the incidence of hurricanes in the Gulf of Mexico or deep depressions in the North Atlantic will make its deep-water operations more hazardous. Such forecasts are entirely dependent on both reliable measurements of how the climate has been changing in recent decades and on whether it is possible to predict future trends. This sets more demanding criteria on what we need to know and whether it is sufficient to make useful forecasts. This applies to all areas of climate variability and climate change. So, the improved understanding of the long-term climatic processes that have shaped the world around us, while central to the making of more efficient use of natural resources, can be dealt with in general terms. The more immediate issues of how the climate could change in the near future and how it will shape our plans for weather-sensitive investments requires more detailed analysis.

This analysis depends on how changes in the Earth's structure and extraterrestrial influences are combined with a huge variety of ways in which the different components of the climate system alter over time and interact with each other. These range from the ever-changing motions of the atmosphere, through the variations in the land surface, including vegetation type, soil moisture levels and snow-cover, to sea-surface temperatures (SSTs), the extent of pack-ice in polar regions, plus the stately motions of the deep-ocean currents which may take over a thousand years to complete a single cycle. Add to this that the prevailing climate combined with the distribution continents controls the amount of nutrients washed into the oceans which affects the oceans' productivity and carbon dioxide levels in the atmosphere and you begin to get an idea of the complexity underlying climate change.

Some of these phenomena are predictable and some are not. For instance the features of the Earth's orbital motion which govern the daily (diurnal) and annual cycles in the weather can be predicted with great precision. Other gravitational effects, such as the longer term cycles in the lunar tides, or the possible effects of the other planets in the Solar System on both the Earth and the Sun, can be calculated with considerable precision, but their influence on the climate is much more speculative. The same applies to even longer term changes in the Earth's orbital parameters, which alter the amount of sunlight falling at different latitudes throughout the year. But, as we will see later, this is the most plausible explanation of the periodic nature of the ice ages, which have engulfed much of the planet during much of the last million years.

By comparison, many of the predictions of how the climate will respond to changes in various parts of the system are far less reliable. Indeed, many of them expose the chaotic nature of the weather and the climate (see Box 1.1), so many aspects of climate change may well prove to be wholly unpredictable. But in studying how the various components of the climate interact with one another it may be possible to establish statistical rules about their behaviour. This might then lead to making useful forecasts of the probability of certain outcomes occurring as well as providing valuable insights into how the system works.

The consequence of the complexity is that from time to time we must enter the worlds of mathematics, statistics and physics. To many people, studying other disciplines, which may be influenced by climate change, this may seem a slightly daunting prospect. But it is a necessary condition to understanding whether the issues raised by the claims of climatologists are of real consequence to your chosen discipline. So, there has to be some mathematics and physics in this book. My objective will be, however, to keep the amount down to a minimum and to do my best to make it as user-friendly as possible.

Armed with these insights the objective of this book will be to steer a judicious course through the essential aspects of climate variability and climate change to show many aspects of our lives are influenced by past and present fluctuations. This will then set the scene for considering the urgent issue of the potential impact of human activities on the climate in the future. The first step in this process is Earth's energy balance.

FURTHER READING

A complete reference list is available at the end of the book but the following is a selection of the best books or articles to follow up particular topics within this chapter. Full details of each reference are to be found in the Bibliography.

IPCC (1990), (1992), (1994) and (1995). In terms of obtaining a comprehensive picture of where the debate on the science of climate change has got to, these are the definitive statement of the consensus view. They are, however, both carefully balanced in their analysis, and exhaustive in their presentation of the competing arguments. As such, they are not an easy read and may appear evasive, if not confusing, until you have got a grip of the basic issues. So they are of greatest value once you have got your bearings clearly established.

Burroughs (1999). A vivid introduction to all aspects of the climate and how it affects our lives. As such it provides useful background reading to help in understanding how both the physical geography of the climate, and climate change, are of real concern to everyone.

CHAPTER TWO

RADIATION AND THE EARTH'S ENERGY BALANCE

And God said, Let there be light: and there was light.

Genesis 1,1

Although the Earth's climate is governed by a wide range of factors, the essential, driving process is the supply of energy from the Sun and what happens to this energy when it hits the Earth. To understand how this works, we must consider the following processes:

  i) the properties of solar radiation and also how the Earth re-radiates energy to space;
  ii) how the Earth's atmosphere and surface absorb or reflect solar energy and also re-radiate energy to space; and
  iii) how all these parameters change throughout the year and on longer timescales.

The consideration of variations over timescales longer than a year has to take account of the fact that many of the most important changes may occur over widely differing periods (i.e. from a few years to hundreds of millions of years). The differing time horizons cannot be specified in advance and so we may find ourselves moving back and forth over this wide range as we progress through the book. The only way to deal with these shifts in timescale is to identify clearly what we are talking about at the time we are doing so.

2.1   SOLAR AND TERRESTRIAL RADIATION

At the simplest level the radiative balance of the Earth can be defined as follows: over time the amount of solar radiation absorbed by the atmosphere and the surface beneath it is equal to the amount of heat radiation emitted by the Earth to space.

2.1.1   Radiation Laws

It is a fundamental physical property of matter that any object not at a temperature of absolute zero (-273.16 °C) transmits energy to its surroundings by radiation (for the purposes of discussing the radiative properties of bodies it is necessary to measure the temperature of the body in relation to absolute zero by defining how many degrees Celsius it is above absolute zero – this value is defined as the body's temperature in degrees Kelvin [K]). This radiation is in the form of electromagnetic waves travelling at the speed of light and requiring no intervening medium. Electromagnetic radiation is characterised by its wavelength which can extend over a spectrum from very short gamma (γ) rays, through X-rays and ultraviolet, to the visible, and on to infrared, microwaves and radiowaves (Fig. 2.1). The wavelength of visible light is in the range 0.4–0.7 µm (1 µm $= 10^{-6}$ m).

Throughout this book electromagnetic radiation will be identified in terms of its wavelength (generally in microns [µm] in the visible and infrared, and millimetres [mm] in the microwave region). There are, however, times when it is more convenient to discuss the frequency of the radiation. The relationship between the wavelength and the frequency is:

$$c = \lambda \nu$$

where λ is the wavelength in metres, ν is the frequency in cycles per second, and c is the velocity of light (3×10^8 m s^{-1}).

A body which absorbs all the radiation and, which at any temperature, emits the maximum possible amount of radiant energy is known as a 'black body'. In practice no actual substance is truly radiatively 'black'. Moreover, a substance does not have to be black in the visible light to behave like a black body at other wavelengths. For example, snow absorbs very little light but is a highly efficient emitter of infrared radiation. But at any given wavelength if a substance is a good absorber then it is a good emitter, while at wavelengths at which it absorbs weakly it also emits weakly.

The wavelength dependence of the absorptivity and emissivity of a gas, liquid or solid is known as its spectrum. Each substance has its own unique spectrum which may have an elaborate wavelength dependence. This means that the radiative properties of the Earth are made up of the spectral characteristics of the constituents of the atmosphere, the oceans and the land surface. These combine in a way that is essential to explaining the behaviour of the global climate.

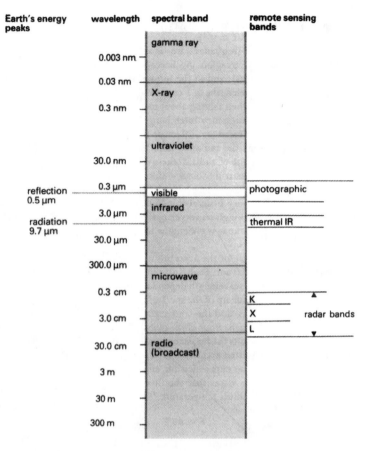

Figure 2.1 The electromagnetic spectrum covers the complete range of wavelengths from the longest radio waves to the shortest γ-rays (Burroughs, 1991, Fig. 2.1).

For a black body, which emits the maximum amount of energy at all wavelengths, the intensity of radiation emitted and the wavelength distribution depend only on the absolute temperature. The expression for emitted radiation is defined by the Stefan–Boltzmann law which states that the flux of radiation from a black body is directly proportional to absolute temperature. That is:

$$F = \sigma T^4$$

where F is the flux of radiation, T is the absolute temperature as measured in K from absolute zero ($-273.16\ °C$), and σ is a constant ($5.670 \times 10^{-7}\ \mathrm{W\,m^{-2}\,(K)^4}$).

It can also be shown that the wavelength at which a black body emits most strongly is inversely proportional to the absolute temperature. Known as the Wien displacement law, this is expressed as:

$$\lambda_m = \alpha/T$$

where λ_m is the wavelength of maximum energy emission in metres and α is a constant ($2.898 \times 10^{-3}\ \mathrm{m\ K}$).

So if the Earth were a black body and the Sun emitted radiation as a black body of temperature 6,000 K, then a relatively simple calculation of the planet's radiation balance produces a figure for the average surface temperature of 270 K. This figure is somewhat lower than the observed value of about 287 K. Furthermore, the Earth does not absorb all the radiation from the Sun and so, in principle, might be expected to be even cooler at around 254 K. The reason for this difference is, as we will see, the properties of the Earth's atmosphere. Widely known as the *Greenhouse Effect* (see Box 2.1), its impact is to alter how the Earth radiates energy from different levels in

BOX 2.1   THE GREENHOUSE EFFECT

Strictly speaking the term the Greenhouse Effect is a misnomer. The principal mechanism operating in a greenhouse is not the trapping of infrared radiation but the restriction of convective losses when air is warmed by contact with ground heated by solar radiation. The radiative properties of the glass in terms of preventing the transmission of terrestrial radiation are inconsequential, as R. W. Wood demonstrated with an elegant experiment in 1909. He showed that when the glass in a model greenhouse was replaced by rock salt, which is transparent to infrared radiation, it makes no difference to the temperature of the air inside. What matters is that the incoming solar radiation is not impeded by the glass, and once the ground and adjacent air is warmed up, turbulent movement of the air does not remove the heat too rapidly.

The fact that greenhouses do not *trap* terrestrial radiation is important not only in understanding how they work, but also in getting an accurate appreciation of how the build-up of radiatively active gases in the atmosphere alters the temperature. Because the density of the atmosphere decreases rapidly with altitude, any absorption of terrestrial radiation will take place principally near the surface (the exception is ozone, which occurs at higher concentrations in the stratosphere [see Box 2.2], and so will exert a different influence on radiative balance). Moreover, since the most important absorber is water vapour, which is concentrated in the lowest levels of the atmosphere, the greatest part of the absorption of terrestrial radiation emitted by the Earth's surface occurs at the bottom of the atmosphere.

If the concentration of radiatively active gases in the atmosphere increases, and the amount of solar energy absorbed at the surface remains unchanged, then the lowest levels of the atmosphere will be warmed up a little. This warming process is the result in the lower atmosphere absorbing and consequently emitting more infrared radiation, both upwards and downwards. In achieving balance between incoming and outgoing radiation this process leads to the surface and lower atmosphere warming and the upper atmosphere cooling (see Fig. B2.1). The consequence of this adjustment to additional absorbing gases is for the surface to warm up and for the

(continued)

atmosphere, as a whole, effectively to radiate from a higher level (i.e. the equivalent black body temperature of the atmosphere required to maintain the Earth's energy balance remains unchanged but rises to a greater altitude). This change is usually expressed in terms of the change in the average net radiation at the top of the troposphere (known as the *tropopause*) and is defined as *radiative forcing*. It is changes in this parameter that is the real measure of the impact of alterations in the greenhouse effect, whether due to natural causes or human activities.

It may seem pedantic to quibble about whether this change can be described as the Greenhouse Effect. After all, the net effect is warming where we all live, just as if we were trapped in a greenhouse with no scope to open ventilation windows, so does it matter? The answer has to be 'yes', because the essence of the process is the theoretical change in the temperature profile of the atmosphere. As we will see this is an essential feature in appreciating the arguments about the physical consequence of

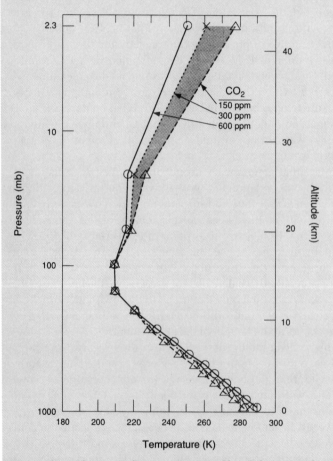

Figure B2.1 The vertical temperature profile of the atmosphere showing how an increase in the concentration of carbon dioxide (CO_2), assuming relative humidity and cloud cover remains unchanged, alters the temperature distribution leading to a warming at low levels and a cooling at high levels (Trenberth, 1992, Fig. 20.4).

the atmospheric build up of radiatively active gases due to human activities. So it is better not to regard these gases as putting the lid on the atmosphere, but rather of subtly altering how it redistributes the heat absorbed from the Sun.

the atmosphere. To understand how this works we need to look more closely at the properties of both solar and terrestrial radiation.

2.1.2 Solar Radiation

The Sun does indeed behave rather like a black body with an effective surface temperature of 6,000 K. This means its maximum emission is in the region of 0.5 μm near the middle of the visible portion of the electromagnetic spectrum (Fig. 2.2). Almost 99% of the Sun's radiation is contained in the short wavelength range 0.15–4 μm. Some 9% fall in the ultraviolet, 45% in the visible and the remainder at longer wavelengths. In the atmosphere much of the shorter ultraviolet is absorbed by oxygen and ozone high in the atmosphere. Water vapour and carbon dioxide (CO_2) absorb much of the infrared radiation beyond 1.5–2 μm at lower levels in the atmosphere. So the spectrum of solar radiation reaching the Earth's surface is cut down to a band from around 0.3 to 2 μm (Fig. 2.2).

2.1.3 Terrestrial Radiation

A black body with a temperature of 287 K emits its energy in the mid-infrared. Most of this energy is emitted in the range 4–50 μm (Fig. 2.3). With the Earth the radiation of heat is complicated by both emissivity of the surface and, even more important, the spectral characteristics of the atmosphere. In practice, most land surfaces and the oceans are efficient emitters of infrared radiation, and so can be regarded as radiatively 'black' at these wavelengths. This means that, at any given point, the amount of outgoing radiation from the Earth is proportional to the fourth power of the temperature, and the characteristics of the intervening atmosphere. Because the principal atmospheric gases (oxygen and nitrogen) do not absorb appreciable amounts of infrared radiation, the radiative properties of the atmosphere are dominated by certain trace gases, notably water vapour, carbon dioxide and ozone. Each of these interacts with infrared radiation in its own way, so the surface radiation is modified by absorption and re-emission in the atmosphere by these trace constituents whose temperatures will generally be different from that of the surface. Because this upwelling energy comes from both the surface and the atmosphere it is defined as *terrestrial radiation*.

The effect of the radiatively active trace gases on the form of the terrestrial radiation is complex. Each species has a unique set of absorption and emission properties known as its *molecular spectrum* which is the product of its molecular structure. This spectrum is made up of a large number of narrow features, referred to as *spectral lines*, which tend to be grouped in broader

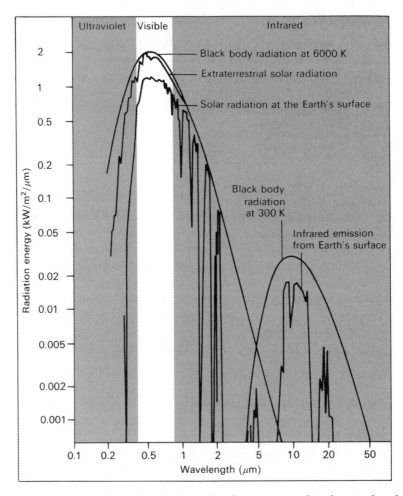

Figure 2.2 The energy from the Sun is largely concentrated in the wavelength 0.2–4 μm (Burroughs, 1991, Fig. 2.2).

bands around certain wavelengths. For different molecules the spectral lines and bands are at different wavelengths. The intensity of individual spectral lines and broader bands varies in a way that is related to the physical properties of each gas. So, we need to know precisely how each gas absorbs and emits infrared radiation to understand how the climate is controlled by variations of the concentration of these gases.

One simple way to quantify the impact of the naturally occurring radiatively active gases is to estimate the contribution they each make to the warming of the Earth above the figure of 254 K. This calculation shows that water vapour contributes 21 K, carbon dioxide (CO_2) 7 K, and ozone (O_3) 2 K. These figures show that water vapour is the most important greenhouse gas. Moreover, if the climate warms, the amount of water vapour in the atmosphere will increase. On its own, this will have a positive feedback (see Box 1.1) on the warming process, while a cooling could be amplified

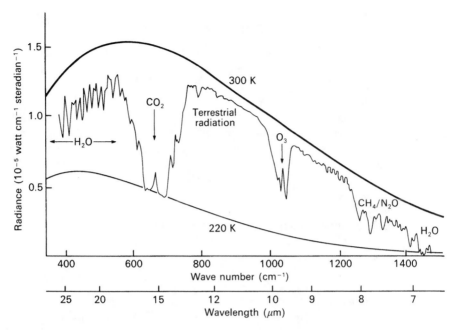

Figure 2.3 The energy emitted by the Earth (terrestrial radiation), as measured by an orbiting satellite, is concentrated on the 4–50 μm region and the amount radiated at any wavelength depends on both the temperature and trace constituents in the atmosphere (Burroughs, 1991, Fig. 2.3).

by the reduction of water vapour in the atmosphere. The significant contribution of CO_2 and O_3 highlight the potential impact of human activities which alter the concentrations of these gases in the atmosphere. In addition, other products of human activities (e.g. methane, oxides of nitrogen, sulphur dioxide and chlorofluorocarbons [CFCs]) will also modify the radiative properties of the atmosphere.

The impact of the most important radiatively active trace constituents shows up clearly in the spectrum of terrestrial radiation observed from space (Fig. 2.3). This spectrum holds the key to understanding how changes in these important trace gases exert control over the climate. Taking the example of the CO_2 band centred around 15 μm, the curve shows that, where this gas absorbs and emits most strongly, the radiation escaping to space comes from high in the atmosphere where the temperature is low (typically ~220 K). Where the atmosphere is transparent on either side of the CO_2 band (often termed *atmospheric window regions*), the radiation comes from the bottom of the atmosphere or the Earth's surface, which is warmer (typically ~287 K). This means that, if the amount of CO_2 in the atmosphere increases, owing to human activities, for example, the band around 15 μm will become more opaque and effectively broaden. This will make no difference where the absorption is strong. But, in other parts of the band the increase in absorption and emission will reduce the amount of terrestrial radiation emitted by CO_2 to space as more of it will come from

BOX 2.2 PHOTOCHEMICAL PROCESSES

A number of minor constituents of the atmosphere are created by the effect of the absorption of short wavelength (ultraviolet [UV]) solar radiation. This high energy radiation breaks down certain molecules in the atmosphere to form highly reactive fragments known as *free radicals* which react with one another and with other molecules in the atmosphere to form new molecular species. The most important absorbers of UV radiation are oxygen (O_2) and ozone (O_3). The former absorbs photons with a wavelength shorter than 240 nm. This absorption process uses the photon's energy to break apart the bonds holding the oxygen atoms together:

$$O_2 + h\nu \rightarrow O + O$$

where $h\nu$ is the energy of a photon having a frequency ν and h is Planck's constant. Oxygen atoms (O) produced are free radicals which react with other oxygen molecules to form ozone,

$$O + O_2 + M \rightarrow O_3 + M$$

where M denotes any air molecule, usually nitrogen or oxygen, that acquires the excess energy generated in this reaction, and dissipates to surrounding molecules by colliding with them, thereby denying the newly formed ozone molecule the excess energy that would cause it to fall apart to O and O_2. Ozone absorbs solar radiation between 240 and 310 nm to revert to O and O_2,

$$O_3 + h\nu \rightarrow O_2 + O$$

The energy from this reaction heats up the adjacent air a little.

The consequence of this set of reactions is that solar radiation between 200 and 310 nm is absorbed in the upper stratosphere, at an altitude of around 25 to 30 km (shorter wavelength solar radiation is absorbed at higher levels). The dynamic balance between the creation and destruction of O_3 means that the concentration of O_3 is a maximum in the stratosphere. Furthermore, the photochemical nature of O_3 production is the reason why other chemical species can alter the balance between the creation and destruction of ozone. In particular, chlorine atoms formed by the breakdown of CFCs in the stratosphere can, in certain circumstances, participate in efficient catalytic reactions with atomic oxygen (O) which destroy ozone. These reactions are a central component of the creation of the *ozone hole* which has appeared over Antarctica each October since the early 1980s.

Other radiatively important photochemical processes are associated with the formation of other free radicals such as the hydroxyl radical (OH), the methyl radical (CH_3) and nitric oxide (NO) formed by the reactions between atomic oxygen and trace atmospheric constituents as follows:

$$O + H_2O \rightarrow OH + OH$$

$$O + CH_4 \rightarrow OH + CH_3$$

$$O + N_2O \rightarrow NO + NO$$

These free radicals are involved in a wide range of chemical reactions with other atmospheric species, especially pollutants such as sulphur dioxide (SO_2), carbon monoxide (CO), oxides of nitrogen (NO_x) and unburnt hydrocarbons (HCs).

In the stratosphere the many reactions are part of the overall balance of ozone and other species. Of particular interest is the formation of nitric acid (HNO_3) which combines in the formation of reactive ice particles over Antarctica to provide the right conditions for destroying ozone more efficiently. By way of contrast, near ground level the cocktail of pollutants, free radicals and bright sunlight produces raised levels of ozone as part of a photochemical smog. Here again the formation of particulates is an integral feature of the complex chemistry involved.

the cold top of the atmosphere. But, for the Earth's energy budget to remain in balance, the amount of terrestrial radiation must remain constant, so the temperature of the lower atmosphere has to rise to compensate for reduced emission in the CO_2 band. This is the basic physical process underlying the Greenhouse Effect (see Box 2.1). It applies to changes in the concentration of all radiatively active gases, and is central to how the temperature of the Earth adjusts to the amount of energy received from the Sun.

The analysis of the radiative impact of greenhouse gases depends on their distribution in the atmosphere. Most trace constituents are relatively uniformly spread both geographically and vertically. Two of the most important greenhouse gases (water vapour and O_3) do, however, have a much more complicated distribution. Because water vapour concentration depends on temperature and the dynamic balance with clouds and precipitation, which makes up the *hydrological cycle*, the amount in the atmosphere varies substantially from place to place. Humidity levels are highest over the tropical oceans and lowest over continental interiors in winter. In addition, the concentration falls off rapidly with altitude. So any analysis of the radiative impact of water vapour has to take full account of the geographical and seasonal nature of these variations.

With O_3 a different set of physical processes are at work. The majority of O_3 is created by the *photochemical* action of sunlight on oxygen in the upper atmosphere (see Box 2.2). This production process depends on the amount of sunlight and so influences the amount of sunlight absorbed by the atmosphere as well as having an impact on the outgoing terrestrial radiation. As such, O_3 is an important example of various photochemical processes in the atmosphere which are an important factor in the radiation balance of the atmosphere. Because of their dependence on the amount of sunlight present, these processes exhibit a marked annual cycle, especially at high latitudes. In the case of O_3 this annual cycle is then modified by transport and photochemical processes. In addition, pollution in urban areas, notably hydrocarbons and oxides of nitrogen from vehicles, can produce the right conditions for the photochemical production of O_3. This is a drawn-out process and so spreads far beyond urban areas. The result is significant widespread increases in O_3 in the lower atmosphere over much of the more populous parts of the world. So the radiative impact of O_3 is concentrated in the stratosphere and close to the surface, and shows marked geographical and seasonal variations.

In considering how human activities may alter the radiative properties of the atmosphere it is normal to present the impact of a specific change. The commonly adopted measure is to calculate the effect of the equivalent of doubling the atmospheric concentration of CO_2 from pre-industrial levels (around 280 parts per million by volume [ppmv]) to, say, 560 ppmv. This change which is defined as radiative forcing is estimated to be about 4 W m^{-2}, is likely to occur principally from the build up from CO_2, but other gases will also make a contribution (Fig. 2.3). To put this radiative impact in context, the amount of energy received in the flux from the Sun is around 1370 W m^{-2} – which equates to 343 W m^{-2} when distributed uniformly over the Earth's surface, whereas the average amount of energy radiated by the Earth is about 240 W m^{-2} after taking account of how much sunlight is reflected back into space (Section 2.1.4).

2.1.4   The Energy Balance of the Earth

How changes in the atmosphere alter the climate can only be discussed in the context of the other factors which control the radiation balance of the Earth. First, there is the fundamental factor of how the Earth's orbit around the Sun and its own rotation about its tilted axis controls the amount of sunlight falling on any part of the globe (Fig. 2.4). These regular changes control

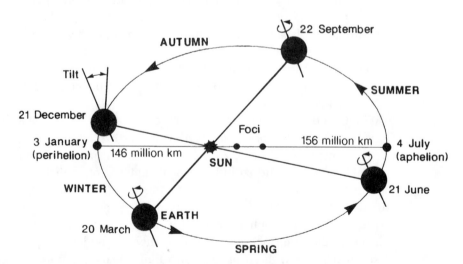

Figure 2.4 The orbit of the Earth is an ellipse but the Sun is not one of the focal points. As a result the Earth is farther from the Sun (at aphelion) at one end of the long axis and closer (at perihelion) at the other. At present the Earth is closest to the Sun in December (Van Andel, 1994, Fig. 5.1).

both the daily (*diurnal*) cycle and the annual seasonal cycle and dominate the climatology of the Earth. The fact that the orbit is elliptical with the Sun at one focus of the ellipse means that the amount of solar radiation reaching the Earth varies throughout the year. At present, the Earth is closest to the Sun in December, so the northern winter is somewhat warmer than it would be if the winter solstice was at the opposite end of the orbit.

For most purposes, orbital variations in the amount of sunlight reaching the Earth during the year can be taken as read and do not need to be considered further. However, longer term variations in orbital parameters do matter. The ratio of the distance of the Sun from the centre of ellipse forming the Earth's orbit and length of the semi-major axis (*eccentricity*) of the orbit changes with time, as does the time of year when the Earth is farthest from the Sun (*precession of the equinoxes*), while the inclination of the Earth's axis to the plane of the orbit (*obliquity*) rocks back and forth periodically. The combination of these effects means that over timescales from several thousand years to a million years the amount of solar energy received at different latitudes at different times of the year changes appreciably.

Having assumed that the output of the Sun is constant, the amount of solar radiation striking the top of the atmosphere at any given latitude and season is fixed by three elements: the eccentricity (e) of the Earth's orbit; the tilt of the Earth's axis to the plane of its orbit – obliquity of the ecliptic (ε); and the longitude of the perihelion of the orbit (ω) with respect to the moving vernal point as the equinoxes precess (Fig. 2.5). Integrated over all latitudes and over an entire year, the energy flux depends only on e. However, the geographic and seasonal pattern of irradiation essentially depends on ε and $e \sin \omega$. The latter is a parameter that describes how the precession of the equinoxes affects the seasonal configuration of Earth–Sun distances. For the purposes of computation the value of $e \sin \omega$ at AD 1950 is subtracted from

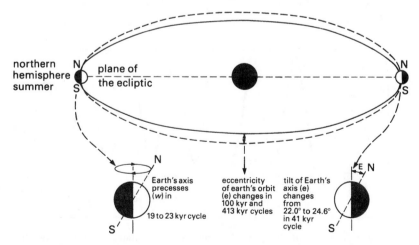

Figure 2.5 The changes in the precession, tilt and shape of the Earth's orbit which are the underlying cause of longer term variations in the climate (Burroughs, 1994, Fig. 6.9).

the value at any other time to give the precession index δ ($e \sin \omega$). This is approximately equal to the deviation from 1950 value of the Earth–Sun distance in June, expressed as a fraction of invariant semi-major axis of the Earth's orbit.

Each of these orbital elements is a quasi-periodic function of time (Fig. 2.6). Although the curves have a large number of harmonic components, much of their variance (see Chapter 7) is dominated by a small number of features. The most important term in eccentricity (e) spectrum has a period of 413 kyr. Eight of the next twelve most significant terms lie in the range from 95 to 136 kyr. In low resolution spectra these terms contribute to a peak that is often loosely referred to as the 100-kyr eccentricity cycle. In contrast, as can be seen in Fig. 2.6, the variation of the obliquity (ε) is a much simpler function and its spectrum is dominated by components with periods near 41 kyr. The precession index (δ) is an intermediate case with its main components being periods near 19 and 23 kyr. In low-resolution spectra these act effectively as a single broad feature with a period of about 22 kyr.

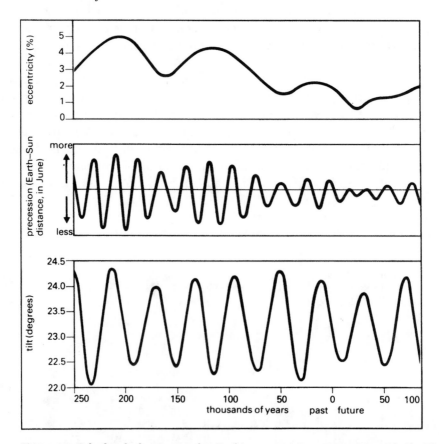

Figure 2.6 Calculated changes in the Earth's eccentricity, precession and tilt. These changes reflect the fact that the Earth's orbit is affected by variations in the gravitational field due to planetary motion (Burroughs, 1994, Fig. 6.10).

Calculations of past and future orbits provide figures for the variation of the orbital elements. The present value of e is 0.017. Over the past million years it has ranged from 0.001 to 0.054. Over the same interval, ε, which is now 23.4°, has ranged from 22.0° to 24.5°, and the precession index (δ, which is defined as zero at AD 1950) has ranged from −6.9% to 3.7%. Because these variations alter the seasonal and latitudinal input of solar radiation to the top of the atmosphere they will affect the climate. Most obviously the changes in the obliquity will have seasonal effects. If the obliquity were reduced to zero, the seasonal cycle would effectively vanish and the pole-to-equator contrasts would sharpen. So low values of the obliquity should correlate with colder periods at high latitudes, which is indeed the case. The eccentricity also exerts a seasonal influence. If e were zero and the Earth had a circular orbit around the Sun, there would be no seasonal effect from this source. The precession of the orbit means that if the summer solstice were shifted towards the perihelion and away from its present position relatively far from the Sun, summers in the northern hemisphere would become warmer and winters colder than they are today.

The key to explaining how these variations can trigger ice ages is the amount of solar radiation received at high latitudes during the summer. This is critical to the growth and decay of ice sheets. At 65° N this quantity has varied by more than 9% during the last 600,000 years. Fluctuations of this order are sufficient to trigger significant changes in the climate which will be explored in detail in Section 8.7. At this stage, however, we need to concentrate on the fact that the total amount of solar energy reaching Earth each year is unaltered by these slow changes in the planet's orbital parameters.

Overall, the total incoming flux of solar radiation is, over time, balanced by the outgoing flux of both solar and terrestrial radiation. This depends on various processes (Fig. 2.7). Some solar radiation is either absorbed or scattered by the atmosphere and the particles and clouds in it. The remainder is either absorbed or reflected by the Earth's surface. The amount of energy absorbed or reflected is dependent on the surface properties. Snow reflects a high proportion of incident sunlight, while moist dark soil is an efficient absorber.

The amount of solar radiation reflected or scattered into space without any change in wavelength is defined as the albedo of the surface. The mean global albedo is about 30%. The albedo of different surfaces can vary from 90% to less than 5%. Examples of the albedo of different surfaces are given in Table 2.1. Roughly half the Earth's surface is obscured by clouds. The albedo of different types and levels of cloud is roughly proportional to cloud thickness. Values for various types of clouds are also given in Table 2.1. As about 30% of incoming solar radiation is reflected or scattered back into space, the remainder must be absorbed. Of this remaining flux, about three quarters (i.e. about half of the total incoming flux) penetrates the atmosphere and is absorbed by the Earth's surface. The remainder

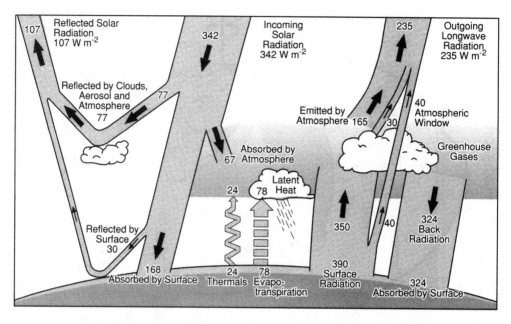

Figure 2.7 The Earth's radiation and energy balance. The net incoming solar radiation of 342 W m^{-2} is partially reflected by clouds and the atmosphere, or at the surface, but 49% is absorbed by the surface. Some of that heat is returned to the atmosphere as sensible heating and considerably more as evapotranspiration that is released as latent heat in precipitation. The amount of energy emitted as thermal infrared radiation from the surface depends on how much is absorbed by the atmosphere which in turn emits radiation both up and down, as part of the greenhouse effect. The net terrestrial radiation lost to space from the surface, from cloud tops and from throughout the atmosphere balances out the total amount of absorbed incoming solar radiation (IPCC, 1995, Fig. 1.3).

(i.e. some 16% of the total incoming flux) is absorbed directly by the atmosphere. Both the atmosphere and the surface re-radiate this absorbed energy as long-wave radiation.

How the solar radiation is absorbed at the Earth's surface differs profoundly on land and at sea. On land most of the energy is absorbed close to the surface, which warms up rapidly, and this increases the amount of terrestrial radiation leaving the surface. At sea the solar radiation penetrates much more deeply, with over 20% reaching a depth of 10 m or more. So the sea is heated to a much greater depth and the surface warms up much more slowly. This means more energy is stored in the top layer of the ocean and less is lost to space as terrestrial radiation. This absorptive capacity of the oceans plays an important part in the dynamics of the Earth's climate. In effect they act as a huge climatic flywheel damping down fluctuations in other parts of the climate.

Clouds play an equally important role. Their high albedo and extensive nature means they are responsible for effectively doubling the albedo of the Earth from the value if there were no clouds. But their transitory nature and variable properties make them particularly difficult to represent accurately

TABLE 2.1 THE PROPORTION OF SUNLIGHT REFLECTED BY DIFFERENT SURFACES (ALBEDO)

Type of Surface	Albedo
Tropical forest	0.10–0.15
Woodland – deciduous	0.15–0.20
Woodland – coniferous	0.05–0.15
Farmland/natural grassland	0.16–0.26
Bare soil	0.05–0.40
Semi-desert/stony desert	0.20–0.30
Sandy desert	0.30–0.45
Tundra	0.18–0.25
Water $(0°–60°)^a$	less than 0.08
Water $(60°–90°)^a$	0.10–1.0
Fresh snow	0.80–0.95
Sea ice	0.25–0.60
Snow-covered vegetation	0.20–0.80
Snow-covered ice	0.75–0.85
Clouds – low	0.60–0.70
Clouds – middle	0.40–0.60
Clouds – high (cirrus)	0.18–0.24
Clouds – cumuliform	0.65–0.75

[a] The closer the sun is to the zenith the more sunlight is absorbed. Also the presence of whitecaps on the surface increases the albedo.

in models of the climate (see Chapter 9), so their role is not easy to quantify accurately. This uncertainty extends to the part they play in modifying the emission of terrestrial radiation. They are efficient absorbers and emitters of infrared radiation. So their thickness and the temperature of their tops plays a great part in how much energy they radiate to space. Furthermore, because water vapour is also the major greenhouse gas, the transition between clouds and vapour poses additional problems in calculating their impact on the outgoing terrestrial radiation. So it is no exaggeration to say that the physics of clouds is the greatest obstacle to improving predictions of climate change.

A related issue is the role of particulates. These may be the product of natural variations (e.g. dust from drought-prone areas) or human activities, including both agriculture and, more important, sulphate particulates from the combustion of fossil fuels. Unlike ice crystals and water droplets in natural clouds, these tiny particles are better reflectors of sunlight than they are absorbers of terrestrial radiation. So their impact on the Earth's energy balance is to reduce the net amount received, and hence to lead to a cooling effect.

This simple input/output analysis is not the whole story. Because the atmosphere and oceans transport energy from one place to another, this motion is an integral part of the Earth's radiative balance. On a global scale the most important feature is that, while most solar energy is absorbed at low latitudes, large amounts of energy is radiated to space, at high latitudes, despite lower temperatures (typically 240 K in polar regions in winter). This process is greatest in winter, when there is little or no solar radiation reaching these regions. This loss must be balanced by energy transport from lower latitudes (Fig. 2.8). Even though this energy transfer is reduced at other times of the year, it provides the energy to fuel the engine which drives the global atmospheric and oceanic circulation throughout the year (see Section 3.6).

Broadly speaking the atmosphere and the oceans transport about the same amount of energy polewards. In the oceans, the Atlantic carries rather more heat northwards than the Pacific (Fig. 2.9). In the Southern hemisphere the Pacific transports roughly twice as much energy southwards as the Indian Ocean, while the Atlantic has a small negative effect carrying energy northwards across the Equator. The role of the oceans in this process cannot be thought as being separate from the atmosphere. There is continual exchange of energy in the form of heat, momentum as winds stir up waves, and moisture in the form of both evaporation from the oceans to atmosphere and precipitation in the opposite direction. This energy flux is particularly great in the vicinity of the major currents, and where cold polar air moves to lower latitudes over warmer water. These regions are the breeding grounds of many of the storm systems which are a feature of mid-latitude weather patterns.

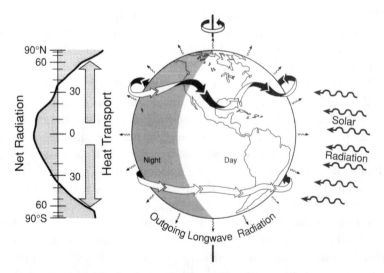

Figure 2.8 In general, the amount of solar energy absorbed by the Earth at each latitude differs from the amount of terrestrial radiation emitted at the same latitude, and so energy has to be transferred from equatorial to polar regions (IPCC, 1995, Fig. 1.2).

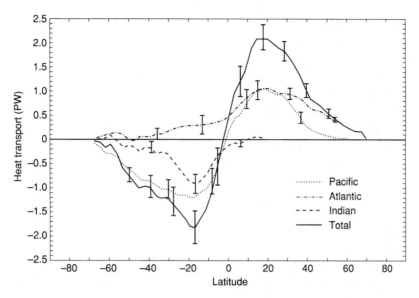

Figure 2.9 The poleward ocean heat transports in each ocean basin, and summer over all oceans shows how the greatest amount of energy is transferred to mid-latitudes. For clarity the poleward transfer in the northern hemisphere is shown as positive, whereas in the southern hemisphere it is shown as negative. So the Atlantic moves rather more energy northwards than the Pacific, whereas the Pacific is the most important transporter southwards, moving about twice as much energy as the Indian Ocean, while the Atlantic is a net transporter towards the Equator. These figures have been calculated using both satellite measurements of outgoing radiation and computer models of the global climate (IPCC, 1995, Fig. 4.6).

Maintaining an energy balance by transport from low to high latitudes has another fundamental implication. This relates to the mean temperatures of different latitude belts. Although these can fluctuate by a certain amount, at any time of the year, overall they will be close to the current climatic normal. Moreover, if one region is experiencing exceptional warmth, adjacent regions at the same latitudes will experience compensating cold conditions. This means that, in the shorter term, climatic fluctuations are principally a matter of regional patterns. These are often the product of the occurrence of certain extremes (e.g. cold winters, hot summers, droughts and excessive rainfall) becoming more frequent than usual. But, in the longer term, it is possible for mean latitudinal temperatures to move to different values while maintaining global energy balance. For the most part, however, these overall changes are small compared with the regional fluctuations which can occur on every timescale.

2.2   SOLAR VARIABILITY

If the amount of energy emitted by the Sun varies over time it is bound to have an effect on the Earth's climate. Since the beginning of the seventeenth century, when Galileo first made telescope observations, it has been known that the Sun exhibits signs of varying activity in the form of sunspots. These darker areas, which are seen at lower latitudes between 30° N and 30° S crossing the face of the Sun as it rotates, are cooler than the surrounding chromosphere. Each sunspot consists of two regions: a dark central *umbra* at a temperature of around 4,000 K and surrounding lighter *penumbra* at around 5,000 K. The darkness is purely a matter of contrast. They appear dark compared to the general brightness of the Sun's brilliant 6,000 K surface temperature. Sunspots vary in their number, size and duration. There may be as many as twenty or thirty spots at any one time, and a single spot may be from 1×10^3 km to 2×10^5 km in diameter with a life cycle from hours to months. The average number of sunspots and their mean area fluctuate over time in a more or less regular manner with a mean period of 11.2 years (Fig. 2.10).

Since 1843, when Heinrich Schwabe discovered that the number of sunspots varied in a regular, predictable manner, the possibility that the energy output of the Sun could vary in a periodic manner has been the subject of great scientific debate. It was not until the late 1980s that satellites provided sufficiently accurate measurements to confirm the output of the Sun did rise and fall during the sunspot cycle (Fig. 2.11). To the surprise of many scientists the output increases with the number of sunspots. It had been assumed that, because the spots are cooler areas, as they multiplied they would reduce the Sun's output. It is now known that the convection associated with the sunspots brings hotter gases from within the Sun to the surface

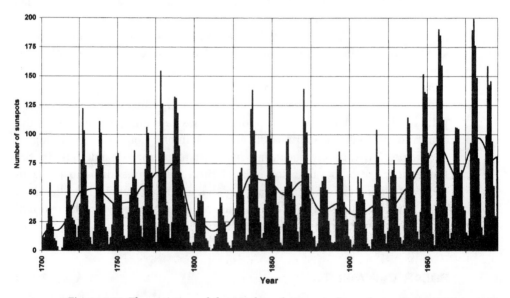

Figure 2.10 The variation of the number of sunspots for each year from 1700 to 1996, together with smoothed data showing longer term fluctuations (Burroughs, 1997, Fig. 5.17).

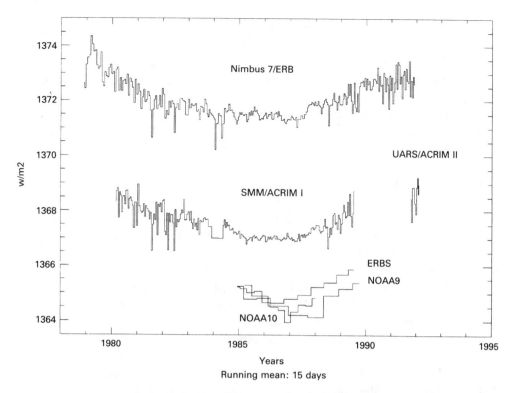

Figure 2.11 Various satellite measurements of the total solar irradiance, showing how the output from the Sun varied between the peaks in sunspot numbers around 1980 and 1990 and the minimum in 1986 (Willson, in Gurney, Foster & Parkinson, 1993, Fig. 1, p. 8).

in the surrounding areas known as *faculae*. These hotter areas outweigh the cooling effect of the sunspots and result in more energy being emitted at the peak of sunspot cycles.

The changes in the total amount of energy emitted by the Sun during the eleven-year sunspot cycle is only about 0.1%. On the basis of energy balance considerations this is not thought sufficient to produce significant climate change. This is, however, only the first part of the story. Much of the change in solar output is concentrated in the UV part of the spectrum. Much of this energy is absorbed high in the atmosphere by both oxygen and ozone molecules (see Box 2.2). If changes at these levels are capable of exerting a disproportionate effect on the weather patterns lower down, then it is possible for small variations in solar radiation to be the cause of larger fluctuations in the climate. Therefore, it is useful to know how solar UV energy affects the upper atmosphere, which is why the creation and destruction of ozone is discussed in Box 2.2. The physical arguments explaining how changes of ozone and other trace constituents in the upper atmosphere may play a part in climate variability and change are then explored in Section 8.5.

On geological timescales the output of the Sun has risen steadily. It is estimated that four billion years ago the solar constant was 80% of the current value and it has risen monotonically since then. These changes are of interest in considering the long-term nature of the response of the global climate to fundamental changes in energy input, but are of no direct consequence for current climate studies.

2.3   SUMMARY

The starting point in examining how the Earth's climate may change over time is to understand the radiative balance of our planet. All the energy that matters comes from the Sun. The incoming flux is matched exactly over time by the outgoing terrestrial energy. This balance is defined by the basic laws of physics. Anything that has the potential to influence this balance, either in global terms, or in respect of one part of the planet as opposed to another, could alter the climate. This means that at every stage of our analysis of fluctuations in the climate, we must anticipate how the amount of solar energy absorbed and reflected by the Earth, together with how much heat is radiated to space by the planet will be affected by the changes we are exploring. In addition, anything that can alter how much solar radiation reaches the Earth (e.g. solar activity, orbital variations or even amounts of cosmic dust) could play a part in altering the climate.

So there are a lot of things that can change in the Earth's energy balance. What is more, they are connected in a variety of ways. These connections are best considered in the context of the various components of the Earth's climate. This means we must now move on to how the balance between incoming and outgoing radiation establishes the global pattern of climatic regimes.

QUESTIONS

1   If the radiative forcing, owing to a doubling in the atmospheric concentration of CO_2, is 4 W m^{-2}, by what amount does the extent of cloud cover have to increase to offset this radiative effect by reflecting additional sunlight back into space? (To do this calculation use the albedo figures in Table 2.1 and assume that at any given time half the Earth's surface is covered by clouds.)

2   Why does the figure calculated in answer to Question 1 overstate the impact of clouds on the radiative balance of the Earth?

3   Why is the radiation budget at any particular place seldom in balance?

FURTHER READING

A complete reference list is available at the end of the book but the following is a selection of the best books or articles to follow up particular topics within this chapter. Full details of each reference are to be found in the Bibliography.

Houghton (1986). This provides the basic physical principles of the behaviour of atmospheres: heavy on mathematics, and not an easy read, but does contain the essence of the physics.

Houghton (1997). A highly accessible discussion of the many issues surrounding the global warming, with a particularly clear section on the physics of the Greenhouse Effect.

IPCC (1994). Provides the definitive analysis of the physical issues involved in the radiative processes governing the Earth's energy balance and how this might be modified by human activities.

McIlveen (1992). A readable and comprehensive presentation of many aspects of meteorology and climate change.

CHAPTER THREE

THE ELEMENTS OF THE CLIMATE

I am the daughter of Earth and Water,
  And the nursling of the Sky;
I pass through the pores of the ocean and shores;
  I change, but I cannot die.

Percy Bysshe Shelley, 1792–1822

In understanding what constitutes climate variability and climate change, and predicting how each may evolve in the future, we need to have a clear idea of which aspects of the climate matter. This is crucial as these fluctuations are not simply a matter of the Earth as a whole warming up or cooling down, or even of certain regional climates (e.g. deserts) expanding or contracting. The mechanics of change may be driven by the physical processes described in Chapter 2, but how they combine with the different components of the climate is complicated by the network of links within the system. So, the objectives of this chapter are to establish what the most important components of the climate are, and to identify the strongest links between them, so we can have a framework for examining the essential aspects of climate variability and climate change.

3.1   THE ATMOSPHERE AND OCEANS IN MOTION

How the atmosphere and the oceans transfer energy around the globe is the key to climate studies. It is also central to the sciences of meteorology and climatology. There are many more-than-adequate books on these

subjects (see Bibliography) and so many of the physical funda
or details of the Earth's climate will not be regurgitated here. Instead
we will concentrate on those latent aspects which are less frequently
addressed but, which are central to understanding the behaviour of the
Earth's climate and how it may change of its own accord or be provoked
into changing.

When viewed from space the atmosphere appears to consist principally of
an endless succession of transient eddies swirling erratically across the face
of the planet. If observed for long enough clear patterns emerge. The most
obvious is the huge pulse of the annual cycle which moves from hemisphere
to hemisphere. If watched with heat-sensing equipment it looks for all the
world like a great blush spreading from one pole to the other – throughout
the course of the year. At the more detailed level the essential features of the
global circulation become obvious (Fig. 3.1). In the equatorial regions the

Figure 3.1 Viewed from a geostationary satellite over the equator many components of
the Earth's climate system can be seen in a single image (with permission of the
Japanese Meteorological Agency; Burroughs, 1991, Fig. 1.1).

planet is girdled by a region of intense convective activity – the Intertropical Convergence Zone (ITCZ). During the northern summer this develops into the more sustained expansion of the monsoon over the Indian subcontinent. Then, during late summer and early autumn, activity in the ITCZ can erupt, now and then, into more organised outbursts as tropical cyclones emerge in the easterly flow and then, as they intensify, swerve northwards before merging into the high-latitude westerly flow.

To the north and south of this tropical hive of activity the continental deserts mark regions of subsiding air and low rainfall. Beyond this, the westerly conveyor belt of mid-latitude depressions swirl endlessly poleward. In the southern hemisphere this procession continues unabated throughout the year with only a modest shift to higher latitude in the austral summer. In the northern hemisphere there is a crescendo of activity in the winter half of the year, with a marked lull during the summer. Polar regions covered with ice and shrouded by clouds conceal much of their behaviour. But, beneath this the Arctic and Antarctic pack ice comes and goes (see Section 3.4), while much of the northern continents is blanketed by snow during the winter: the extent of which varies markedly from year to year.

Associated vertical motion of all these systems is central to understanding the climatic processes at work. These are shown in Fig. 3.2. The rising air in equatorial regions generates the ITCZ. When it reaches an altitude of around

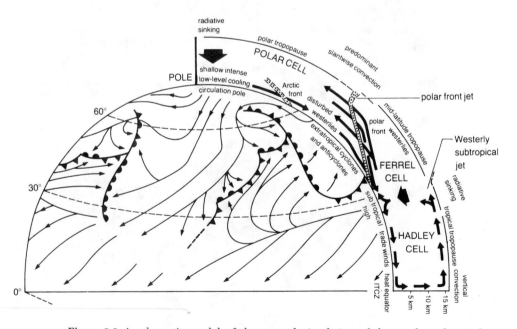

Figure 3.2 A schematic model of the general circulation of the northern hemisphere (in cross-section) the locations of both the principal surface features, the vertical circulation and the polar front jet stream and the westerly sub-tropical jet (Musk, 1988, Fig. 12.5).

12 to 15 km virtually all the moisture has been wrung out of it, and it spreads out and then descends to the north and south of this region creating zones of dry, hot air which maintains the deserts in these parts of the world and also drives the Trade winds which flow back towards the equator. First interpreted by George Hadley in 1735, this circulation pattern now bears his name (*the Hadley Cell*). The mid-latitude depressions also raise huge quantities of air as they swirl polewards. This air descends over polar regions and creates cold deserts in these regions.

The movement of the oceans is not so easy to see but is just as important. Satellite measurements do, however, confirm the picture of ocean currents built up from centuries of maritime observations (Fig. 3.3). They also show that the flow is complicated to eddies and swirls (Fig. 3.4) which mirror the turbulent nature of the atmosphere. In addition, temperature measurements show that, on top of the annual pulse, there are major fluctuations in sea surface temperatures (SSTs) on longer timescales, especially in the equatorial Pacific (see Section 3.5). Finally, completely hidden from view the deep waters of the oceans silently circulate, moving huge amounts of energy on the timescale of centuries.

All of these processes can interact with one another to produce variations on every possible timescale. In the short term they represent the fascinating extremes of our daily weather. But where they combine in a more sustained way then they produce the recipe for natural variations of the climate. So, the first step is to appreciate what drives these motions, and why they may fluctuate of their own accord. Then we can move on to how they can act in concert.

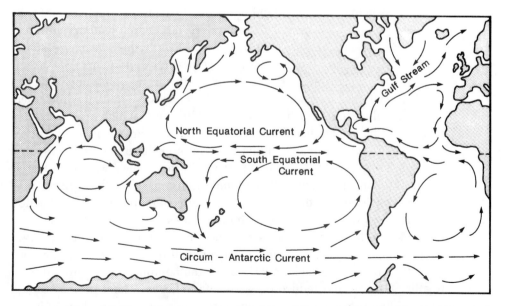

Figure 3.3 The major ocean current systems of the world (Van Andel, 1994, Fig. 10.2).

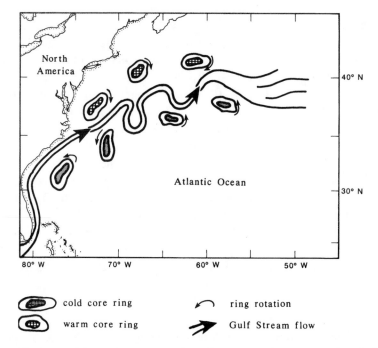

North
America

Atlantic Ocean

40° N

30° N

80° W   70° W   60° W   50° W

cold core ring   ring rotation

warm core ring   Gulf Stream flow

Figure 3.4 Schematic representation of ring structures associated with the Gulf Stream, which are typical of the detailed structure that exists in the major ocean currents (Bryant, 1997, Fig. 4.6).

3.2   ATMOSPHERIC CIRCULATION PATTERNS

The first issue to be addressed in finding out the causes of fluctuations in the climate is why large-scale global atmospheric circulation patterns shift from year to year. This is not a matter of looking at individual weather systems, but instead concentrating on their average positions and paths they take. If mid-latitude depressions take a different path during one winter, or for a number of winters, then some regions will get excessive rainfall while others suffer drought. Similarly, the repeated establishment of high pressure where more variable weather usually occurs can produce abnormally cold winters or hot summers. So understanding the nature of these patterns is the first stage of our search. In the case of patterns in the tropics, which are just as important, and in particular the summer monsoon in southern Asia, these are best considered in the context of sea surface temperatures (see Section 3.5). Here, we will concentrate on mid-latitudes.

The key to long-lasting weather patterns is found in the middle levels of the atmosphere. By showing the average height of a given pressure surface (e.g. 500 mb), it is possible to strip away the complication of ground-level weather features and reduce the associated effects of land masses. So the winter 500-mb northern hemisphere circulation pattern (see Fig. 3.5(a)) is dominated by an extensive asymmetric cyclonic vortex with the primary centre over the eastern Canadian arctic and a secondary one over eastern Siberia. In summer (see Fig. 3.5(b)) the pattern is similar, but the vortex is much less pronounced. These deviations from a purely circular pattern represent troughs and ridges in the circulation. At any time the number of troughs and ridges may vary as will their position and amplitude. These patterns are known as *long waves* (or *Rossby waves* – after the Swedish meteorologist Carl Gustav Rossby who first provided a physical explanation of their origin).

The basic physical explanation of the existence of long waves in the upper atmosphere is linked to the circulation of transient eddies at lower levels. The combined spin of one of these low pressure systems and its rotation about the Earth's axis (known as *vorticity*) effectively remains constant. So, a system in a stream of air moving towards polar regions will tend to lose vorticity as the distance from the Earth's axis of rotation reduces. This is compensated by the flow adopting an anticyclonic curvature (clockwise in the northern hemisphere) relative to the surface at higher latitude. Eventually, the curved path takes the airstream back towards the equator. As it reaches lower latitudes the reverse effect will take place as the distance from the axis of rotation increases and the airflow will adopt a cyclonic curvature. So the airstream tends to swing back and forth as it circulates the Earth. The climatological mean of this flow has two major troughs at around 70° W and 150° W. Their position is linked to the impact of the major mountain ranges on the flow, notably the Tibetan plateau and the Rocky Mountains, together with the heat sources such as ocean currents (in winter) and land masses (in summer). Because the southern hemisphere is largely covered by water, the pattern is much more symmetrical and shows less variation from winter to summer (see Fig. 3.6(a) and (b)).

What is not evident from these broad patterns is the nature of winds in the upper atmosphere. The contour maps record the thickness of the atmosphere between sea level and the 700-mb surface, and this thickness is proportional to the mean temperature – low thickness values correspond to cold air and high thickness values to warm air. This means that the circumpolar vortices (see Figs. 3.5 and 3.6) reflect a poleward decrease in temperature and this drives strong westerly winds in the upper atmosphere. But, while the thickness decreases gradually with increasing latitude, the strongest winds are concentrated in a narrow region, often situated around 30° of latitude at an altitude between 9 and 14 km. The concentrated wind flow is known as the *jet stream*, and reaches maximum speeds of 160 to 240 kph, and can exceed 450 kph in winter. The mainstream core is associated with the principal troughs of the Rossby long waves. The movement of surface weather systems is governed by this circulation. The reason for the restriction of the winds to a narrow core is not fully understood. Moreover, the structure can be complicated, especially in the northern

(a)

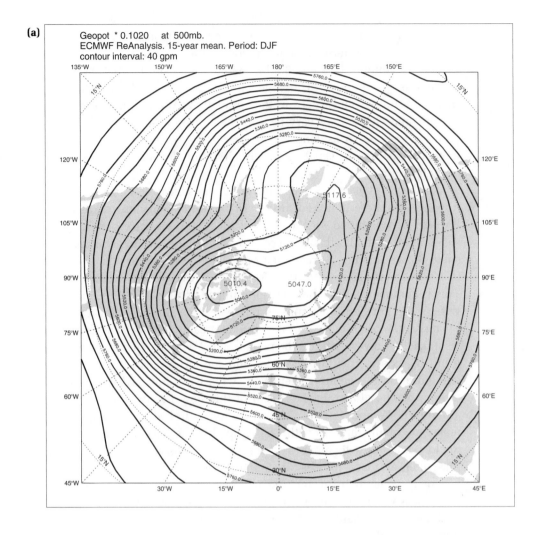

Geopot * 0.1020   at 500mb.
ECMWF ReAnalysis. 15-year mean. Period: DJF
contour interval: 40 gpm

hemisphere winter when the jet stream often has two branches (the subtro-
pical and polar front jet streams).

Circulation patterns, which are radically different from those in Figs. 3.5
and 3.6, can last a month or two. They occur irregularly but are more
pronounced in the winter when the circulation is strongest. Ridges and
troughs can become accentuated, adopt different positions and even split
up into cellular patterns. An extreme example of such a pattern occurred in
the winter of 1962–1963. Figure 3.7 shows the mean height of the 700-mb
surface in January 1963. The striking features are a pronounced ridge off
the west coast of the United States (40° N, 125° W) and a well-defined
anticyclonic cell just to the south of Iceland (60° N, 15° W). This set the
stage for the extreme weather. Cold arctic air was drawn down into the
central United States and into Europe. Conversely, warm tropical air was
drawn far north to Alaska and western Greenland. The net effect was that
while an extreme negative anomaly of −10 °C for the month was observed

(b)

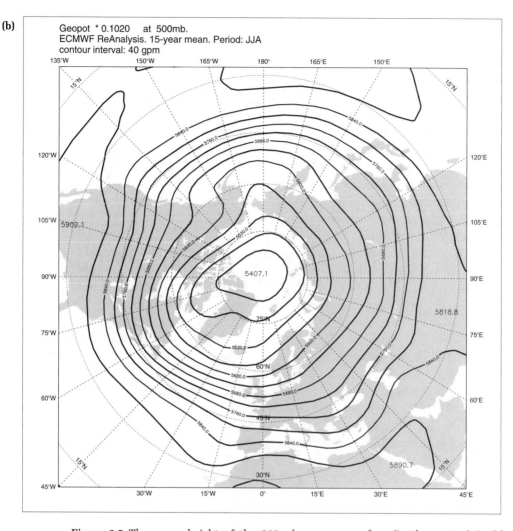

Figure 3.5 The mean height of the 500 mb pressure surface (in decametres) in **(a)** winter and **(b)** summer for the northern hemisphere (with permission of the ECMWF).

in Poland, an equal positive departure occurred over western Greenland (see Fig. 3.8), and the mean hemispheric temperature was close to normal (see Section 2.1.4).

Such an extreme pattern, which is usually termed '*blocking*', shows how important it is to understand the causes of anomalous atmospheric behaviour. Changes in the incidence of extreme seasons, like the winter of 1962–1963, are a major factor in variations of the climate during recent centuries in Europe (see Section 4.9). Therefore, it is important to know what defines their occurrence as opposed to stronger, more symmetric circulation patterns which bring much milder winters to Europe and the eastern United States. Thus, we need to address the basic questions of what causes the number, amplitude and position of the Rossby waves to change and then to remain stuck in a given pattern for weeks or months.

(a)

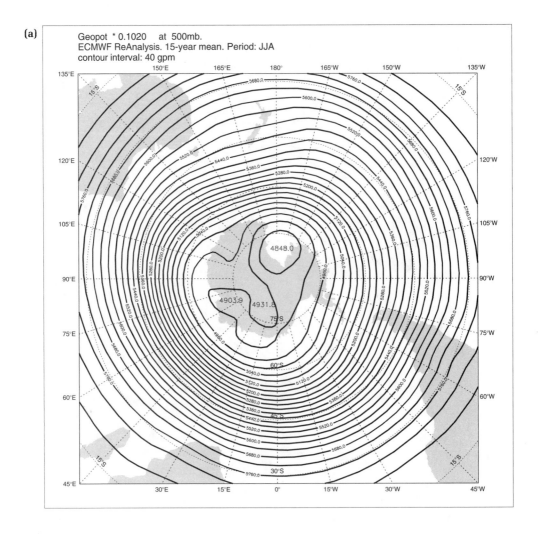

Geopot * 0.1020   at 500mb.
ECMWF ReAnalysis. 15-year mean. Period: JJA
contour interval: 40 gpm

As noted earlier, in the northern hemisphere the distribution of land masses and the major mountain ranges plays a major role, but this does not explain why the number of waves around the globe may range from three to six or why they can vary from only small ripples on a strong circumpolar vortex to exaggerated meanderings with isolated cells. It follows from the description of the requirement to conserve vorticity in the high level flow that an important factor is the speed of the upper atmosphere westerlies. There is some evidence that when the wind speeds assume a critical value this enhances the chances of strong standing waves building up downstream from the troughs at 70° W and 150° E. But this begs the question of what governs the changing speed of the winds from year to year and within seasons.

Any explanation of what causes different circulation patterns must also show why their incidence changes. In winter in the northern hemisphere just four or five categories of circulation regimes occur about three-quarters

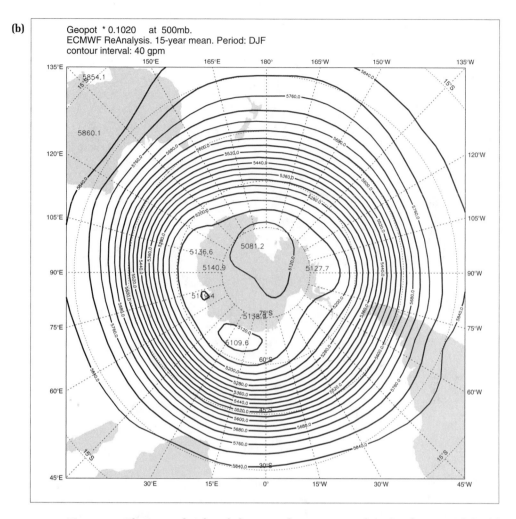

Figure 3.6 The mean height of the 500 mb pressure surface (in decametres) in **(a)** winter and **(b)** summer for the southern hemisphere (with permission of the ECMWF).

of the time. This fact may hold the key to understanding what controls the switches between regimes, as the limited number of states suggests there are tight constraints on which patterns are permissible in strong circulation systems. Any insight may then be extended to other times of the year and other parts of the world where circulation regimes exercise less influence on the climate. So, the essential issues are why a given regime becomes established, what causes it to shift suddenly to a different form and, above all, what defines the incidence of given regimes.

The obvious place to look for controlling influences is in the slowly varying components of the climate system. This means, first and foremost, the oceans, but could also involve the extent of snow and ice in high latitudes. But before doing so, the analysis must be set in the context of

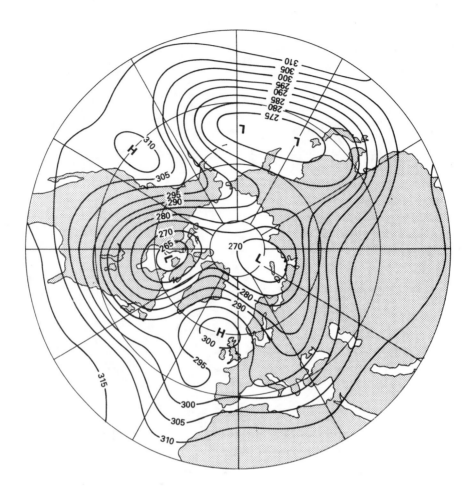

Figure 3.7 The mean height of the 700-mb pressure surface (in decametres) for January 1963, showing the pronounced wave pattern in mid-latitudes due to 'blocking' off the west coast of the United States and close to the British Isles (Burroughs, 1994, Fig. 5.3).

the physical processes outlined in Chapter 2. The atmosphere is driven principally by the solar energy absorbed at the surface at low latitudes, and the terrestrial energy radiated at higher latitudes (see Fig. 2.5). So in explaining how the meanderings of the jet stream in the upper atmosphere may lead to variations in the climate, we must not fall into the trap of assuming that small changes at high altitudes can exert a disproportionate impact at the surface. The fundamental point is that the weather is driven by the energy available at the bottom of the atmosphere. Although the upper atmosphere winds appear to be steering the progress of the surface weather, they are essentially the product of the amount and distribution of the solar energy absorbed into the atmospheric heat engine. But, the devil is in the detail: in explaining changes in climatic patterns there are all manner of subtle alterations of the radiation and water balance which

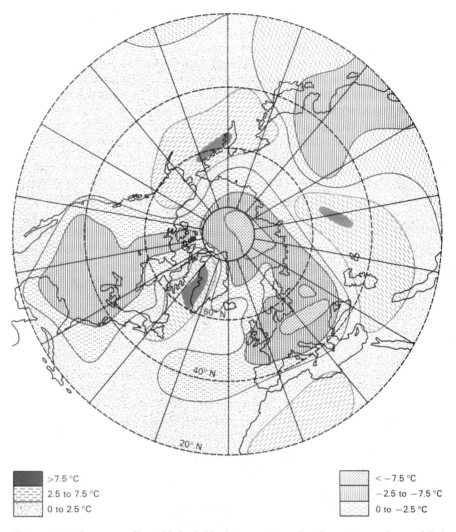

>7.5 °C	< −7.5 °C
2.5 to 7.5 °C	−2.5 to −7.5 °C
0 to 2.5 °C	0 to −2.5 °C

Figure 3.8 When a well-established blocking pattern develops, it can have global implications. The winter of December 1962 to February 1963 is an extreme example of such a global circulation anomaly. This diagram shows that as a consequence of this pattern adjacent regions of the mid-latitudes of the northern hemisphere experienced compensating extremes with northern Europe, the United States and Japan having well below normal temperatures, and Greenland, central Asia and Alaska being well above average (Burroughs, 1991, Fig. 12.4).

may exert longer term influences, some of which may occur in the upper atmosphere.

There is one other aspect of upper atmosphere winds which needs to be mentioned here. This is the periodic reversal of the winds in the lower stratosphere at levels between around 20 and 30 km over equatorial regions. These winds go through a cycle every twenty-seven months or so from being strongly easterly to nearly as strongly westerly (Fig. 3.9). The wind regime propagates downwards with the reversal starting at higher levels. This behav-

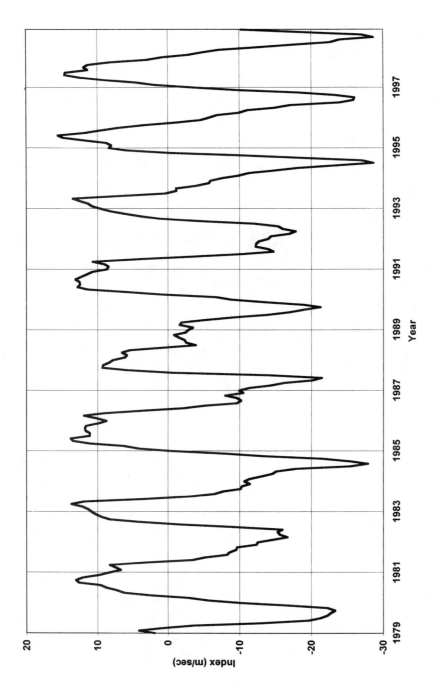

Figure 3.9 An index of the stratospheric equatorial winds at the 30 mb-level (altitude of 24 km) showing the quasi-biennial oscillation (QBO) in the wind direction (data from NOAA).

iour is known as the Quasi-Biennial Oscillation (QBO) and appears to have weak echoes in many features of the weather at lower levels, including many temperature records, the behaviour of the El Niño Southern Oscillation (see Section 3.7), and the incidence of hurricane activity in the Atlantic. It is also the only clearly defined cycle in the weather longer than the dominant annual cycle.

What is particularly relevant about the QBO is that it is the product of internal variability of the atmosphere. The accepted physical explanation involves a combination of processes in the dissipation of upward-propagating easterly Kelvin waves (see Section 3.7) and westerly Rossby waves in the stratosphere. These waves originate in the troposphere and lose momentum in the stratosphere by a process of radiative damping. This involves rising air cooling and radiating less strongly than the air that is warmed in the descending part of the wave pattern. Westerly momentum is imparted by the decaying Kelvin waves in the shear zone beneath the downward-propagating westerly phase of the QBO. Rossby waves perform a similar function in respect of the easterly phase. The fascinating feature of this process is that these two types of wave combine to produce such a regular reversal of the upper atmosphere winds.

3.3 RADIATION BALANCE

As described in Chapter 2, the proportion of incoming solar radiation absorbed by the Earth depends on the absorption, reflection and scattering properties of the atmosphere and the surface. Satellite measurements have now provided a global picture of these processes and confirm the broad figures quoted in Section 2.1.4 and Table 2.1. Where there are no clouds the oceans are the darkest regions of the globe. They have albedo values ranging from 6% to 10% in the low latitudes and 15% to 20% near the poles. Ocean albedo increases at high latitude because at low sun angles water reflects sunlight more effectively (see Table 2.1). The brightest parts of the globe are the snow-covered Arctic and Antarctic which can reflect over 80% of the incident sunlight. The next brightest areas are the major deserts. The Sahara and the Saudi Arabian desert reflect as much as 40% of the incident solar radiation. The other major deserts (the Gobi and the Gibson) reflect about 25% to 30%. By comparison, the tropical rain forests of South America and Central Africa, as the darkest land surfaces, have albedos from 10% to 15%.

The pattern of outgoing long-wave radiation is more systematic. This reflects the fact that the temperature of the surface and the atmosphere decreases relatively uniformly from the equator to the poles. In effect, these surfaces are close to being black bodies when it comes to emitting long-wave radiation and the emissivity of various surfaces does not vary

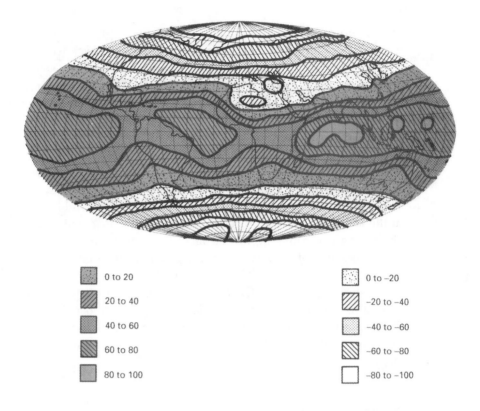

	0 to 20		0 to −20
	20 to 40		−20 to −40
	40 to 60		−40 to −60
	60 to 80		−60 to −80
	80 to 100		−80 to −100

Figure 3.10 Satellite measurements over the years show how the net balance between incoming solar radiation and outgoing terrestrial radiation varies with latitude and longitude. While these observations show considerable variations between different parts of the tropics and subtropics, the important feature is that energy is pumped into the climate system at low latitudes and escapes at high latitudes (Burroughs, 1991, Fig. 13.4).

appreciably. The average amount of energy radiated to space decreases from a maximum of 330 W m^{-2} in the tropics to about 150 W m^{-2} in the polar regions. When combined with the figures for the amount of solar radiation absorbed at the surface and in the atmosphere, a general pattern of the net heating of the atmosphere at low latitudes and cooling at high latitudes emerges from these measurements.

Before reviewing the impact of clouds on the energy balance both globally and regionally, it is necessary to consider the implications of the clear sky albedo measurements for long-term fluctuations in the weather. The most highly reflecting areas are also among the most variable climatic elements. For instance, increases in the extent of winter snow and ice cover in polar regions will have a significant effect on the amount of sunlight reflected into space. This on its own will produce a cooling effect so, the longer anomalous cover lasts, the greater the impact will last. More complicated is the effect of forests across the northern continents. As shown in Table 2.1, the albedo of snow-covered areas varies considerably. The average figure for forests is sub-

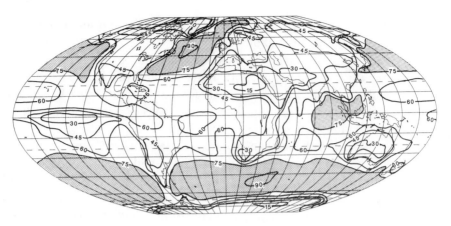

Figure 3.11 Global distribution of annual mean cloud amount expressed as a percent cover. Areas greater than 75% are shaded (Bryant, 1997, Fig. 2.10).

stantially less than for open country (around 0.5 as opposed to nearer 0.7), so the removal of forests at high latitudes would increase the albedo in winter and spring and have an appreciable cooling effect.

A similar but somewhat more surprising effect relates to an expansion of the major deserts of the world. Here again the change will lead to an increase in albedo (see Table 2.1) and hence to more solar radiation being reflected into space. This effect can be seen in Fig. 3.10 as the region of net cooling over the Sahara and the Middle East. So while deserts are regarded as hot places their expansion could lead to a general cooling, unless associated with some compensating changes in cloudiness. This means that if either natural climatic variability or human activities lead to an increase in the world's deserts this is likely to have a cooling effect. So, if global warming leads to expanding deserts, any consequent cooling will be a negative feedback mechanism tending to damp down the warming trend.

The global distribution of clouds shows that they are most common over the mid-latitude storm tracks of both hemispheres (Fig. 3.11). They also have a less striking maximum over the tropics, especially the region around southeast Asia. On average, about 60% of the Earth is covered by clouds at any given time. Clouds are almost always more reflective than the ocean surface and the land except where there is snow. So when clouds are present they reflect more solar energy into space than do areas which have clear skies. Overall, their effect is approximately to double the albedo of the planet from what it would be in the absence of clouds to a value of about 30% (see Section 2.1.4).

The overall effect of clouds depends, however, on their net effect on both incoming and outgoing radiation. Where clouds are present less thermal energy is radiated to space than where the skies are clear. It is the net difference between these two effects which establishes whether the presence of clouds cools or heats the planet. The scale of the blanket-like warming effect depends on the thickness of the clouds and the temperature of their tops. High clouds radiate less than low clouds, and thick clouds are more efficient radiators than thin clouds. The average figures for clouds globally is to reduce the amount of solar radiation absorbed by the Earth by 48 W m^{-2} and to reduce the outgoing heat radiation to space by 31 W m^{-2}. So satellite measurements confirm that clouds have a net cooling effect on the global climate.

While the global effect of clouds is clear, their role in regional climatology and in feedback mechanisms associated with climate change is much more difficult to discern. The blanketing effect of clouds reaches peak values over tropical regions and decreases towards the poles. This is principally because clouds rise to a greater height in the tropics, and the cold tops of deep clouds radiate far less energy than shallower, warmer clouds. So where there are extensive decks of high cirrus clouds, the amount of energy radiated to space, compared with clear skies in the same regions, is reduced by 50 to 100 W m^{-2}. These thick high clouds occur in three main regions. The first is tropical Pacific and Indian Oceans around Indonesia and in the Pacific north of the equator where rising air forms a zone of towering cumulus clouds. The second is the monsoon region of Central Africa and the region of deep convective activity over the northern third of South America. The third is the mid-latitude storm tracks of the North Pacific and North Atlantic Oceans.

The pattern of increased albedo due to clouds is different. The regions associated with the tropical monsoon and deep convective activity reflect large amounts of solar radiation, often exceeding 100 W m^{-2}, as do the clouds associated with the mid-latitude storm tracks in both hemispheres and the extensive stratus decks over the colder oceans. The important difference is that these clouds at high latitudes have less impact on the outgoing thermal radiation as the underlying surface is colder and hence emits less energy whether or not there are clouds. So in the tropics the net effect of clouds is effectively balanced out, but over the mid- and high-latitude oceans polewards of 30° in both hemispheres, clouds have a cooling effect. This negative effect is particularly large over the North Pacific and North Atlantic where it can be between 50 and 100 W m^{-2}.

These are only the most obvious effects of clouds but they provide a clear guide to the complex connections that can lead to apparently periodic climatic fluctuations. The feedback mechanisms that can operate are significant. For instance, if the storm track across the North Atlantic were to move south, as appears to have been the case during the Little Ice Age (see Section 4.9), this could have a significant cooling effect. Taking an extreme example, if the region of strongest cloud forcing at around 45° N underwent a shift southwards to 35° N throughout the year, it could induce a hemispherical average radiative cooling of roughly 3 W m^{-2}. The significance of this figure is that it is comparable to the estimated 4 W m^{-2} radiative forcing arising from the equivalent of doubling of the carbon dioxide in the atmosphere (see

Section 2.1.3). So, although this example may be excessive, the message is clear – sustained natural changes in the distribution of cloud cover could have significant climatic impact.

The satellite records of global cloudiness are the subject of scientific argument. Although weather satellites have been collecting regular pictures of clouds since the early 1960s, which might be expected to provide a continuous record of global cloudiness, this is not the case. Although Russian studies suggest there have been significant variations in cloud cover, other analyses conclude that instrumental limitations throw these conclusions into doubt. Problems arise from the optical properties of different types of clouds and the changing sensitivity of satellite equipment during its lifetime. These issues have not been resolved, but what is clear is that if global cloudiness has changed as much as the Russian measurements suggest they would have had a significant impact on the climate.

3.4   THE HYDROLOGICAL CYCLE

In placing so much emphasis on the radiative balance in understanding the elements of climatic fluctuations, we should not lose sight of how this is interlinked with the behaviour of the hydrological cycle. Water is the most abundant liquid on earth. The total amount with the earth–atmosphere system is estimated to be some $1,384 \times 10^6$ km^3; of this amount 97.2% is contained in the oceans, 0.6% is ground water, 0.02% is contained within rivers and lakes, 2.1% is frozen in ice caps and glaciers (the *cryosphere*) and only 0.001% is in the atmosphere. Nevertheless, water vapour is by far the most important variable constituent of the atmosphere, with a distribution that varies in both time and space. In addition, on timescales of decades and longer, the proportion of water locked up in the cryosphere can change appreciably with significant consequences for the Earth's climate.

At any one time the mean water content of the atmosphere is sufficient to produce a uniform cover of some 25 mm of precipitation over the globe – equivalent to some ten days of rainfall. This means that there is a continual recycling of water between the oceans, the land and the atmosphere, which is known as the *hydrological cycle*. Moreover, because water has such a high latent heat of fusion and evaporation, the phase transitions between ice and liquid water, and between liquid water and water vapour, involve large amounts of energy. So the process of evaporating large amounts of water into the atmosphere, and its subsequent precipitation as rain or snow, is a major factor in the energy transport of the climate. The basic elements of this cycle are well understood and fully described in many meteorological texts. Here we are concerned with the less well understood aspects of how various components of the cycle can vary and their implications for the climate.

Important climatic points about the potential variability of the hydrological cycle relate principally to how much water passes into the atmosphere, the nature and form of the clouds it forms, and how quickly it is precipitated out again. Over the oceans the most important factor is the temperature of the surface, with wind speed playing a secondary role. Over the land the level of soil moisture is a controlling factor, but the presence of living matter complicates the issue considerably (see next section). The amount of water vapour passing into the atmosphere is a combination of evaporation from the soil and transpiration from plants (this combination is defined as *evapotranspiration*), and depends on soil moisture, the air temperature, the temperature of the soil which is related to the amount of sunlight absorbed by the soil, and wind speed. As noted earlier, the levels of water vapour in the atmosphere effect its radiative properties, while its condensation to form clouds is a critical factor in defining the climate.

3.5   THE BIOSPHERE

The involvement of plants on land in the hydrological cycle is only part of the story. The totality of living matter, both on land and in the oceans (the *biosphere*), has the capacity to influence the climate in a variety of ways. The most important is in terms of the control it exerts over certain greenhouse gases. In particular, through photosynthesis the biosphere acts as the fundamental control over the level of CO_2 in the atmosphere. Similarly, the production of CH_4 by the anaerobic decay of vegetation maintains the natural level of this important greenhouse gas (see Section 2.1.3). The totality of the biosphere is not an easy concept to grasp. In climatic terms it is most obvious in how it alters the CO_2 content of the atmosphere throughout the year (see Fig. 8.10). During the northern hemisphere growing season the biosphere draws down CO_2 from the atmosphere only to release much of it during the winter.

The potential of the biosphere to absorb additional CO_2 is an important factor in considering climate change. It is a temporary sink for a proportion of any additional CO_2 injected into the atmosphere. This sequestration is central to any discussion of the climatic implications of past variations in concentrations owing to shifts in the overall productivity of the biosphere, or emissions of this gas into the atmosphere as a result of human activities. Because the productivity of the biosphere rises with increasing CO_2 levels, this slows down the build-up in the atmosphere. This negative feedback mechanism has exerted a major impact on past climates, and is a factor delaying some of the consequences of emissions of this greenhouse gas due to the combustion of fossil fuels.

The formation of particulates as a result of activity in the biosphere is another potentially important aspect of climate change. On land the emission of many volatile organic substances (e.g. terpenes from fir trees) leads

two (Fig. 3.12). More important, there is little evidence that large anomalies are sustained for more than a few months (Fig. 3.13).

In the longer term the satellite observations show that the extent of snow cover declined appreciably between the late 1970s and around 1990. Since then, it has, if anything, increased a little (see Fig. 3.13), while remaining well below the levels of the 1970s and early 1980s. The overall trend correlates closely with changes in the average temperature at higher latitudes of the northern hemisphere. These observations provide some support for the view that reduced snow cover is amplifying the current global warming, but the close links with wider circulation patterns suggest they are driven principally by longer term factors. So they will only contribute significantly to these longer term shifts in the climate if they reinforce the impact of other forcing factors. For example, if the extent of spring and autumn snow cover responds markedly to changes in the amount of sunlight falling at high latitudes during the summer, owing to orbital variations, it could act as a positive feedback mechanism and have a major climatic impact (see Section 2.1.4).

At a regional level, changes in snow cover may exert a greater influence. It has been proposed that the extensive snow cover across Europe during the exceptional winter of 1962–1963 (see Section 3.2) could have been partially instrumental in sustaining the cold weather. This winter was the coldest since 1830 and resulted in virtually all of Europe north of the Alps being covered in deep snow throughout January and February 1963. The prolonged snow cover helped sustain the high pressure region over Scandinavia, which was the principal feature of the abnormal weather patterns. Similarly, in the United States the extensive snow cover that built up during the record-breaking cold spell in December 1983 helped prolong the wintry weather. During January 1984 it is estimated that in parts of the midwest the daytime maximum temperature was 5 °C lower than would have been expected on the basis of the prevailing atmospheric conditions. But, for the reasons noted above, these regional effects do not amount to sufficient reason to attribute longer term climate change to being driven by snow cover variations.

The consequences of changes in the extent of polar pack ice are similar. In some sectors of the northern hemisphere, notably the North Atlantic, the changes in ice cover can affect weather patterns (Fig. 3.14). On average the area of arctic pack ice ranges from a minimum of around seven million square kilometres in late summer to a maximum of some fifteen million square kilometres in early spring. The scale of these changes is, however, small compared with the variations in snow cover, which reach a maximum extent of around forty-five million square kilometres in late winter and decline to less than five million square kilometres in late summer. In the southern hemisphere the reverse is true. Because the Antarctic snow cover is permanent, and winter snow in South America, Australia and New Zealand is small, the most important variations are associated with the extent of

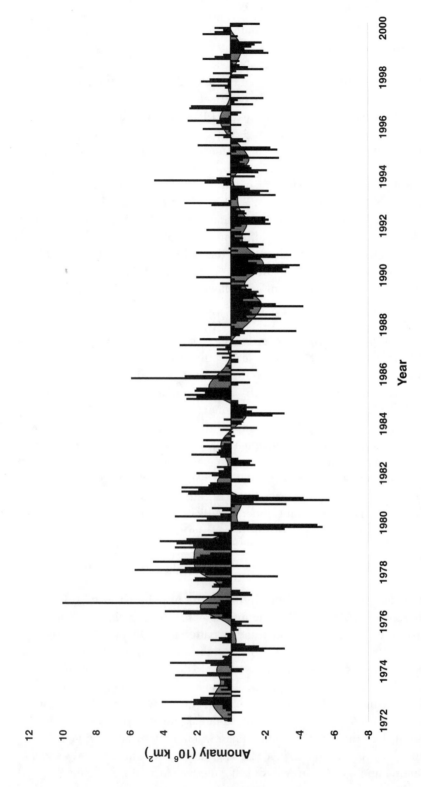

Figure 3.13 Changes in snow cover in northern hemisphere between January 1973 and March 2000 (vertical bars and heavy line) obtained from satellite observations. These fluctuate dramatically from month to month and the long-term decline in snow cover is closely correlated with the rise in temperature in the northern hemisphere north of 30° N (data provided by NOAA).

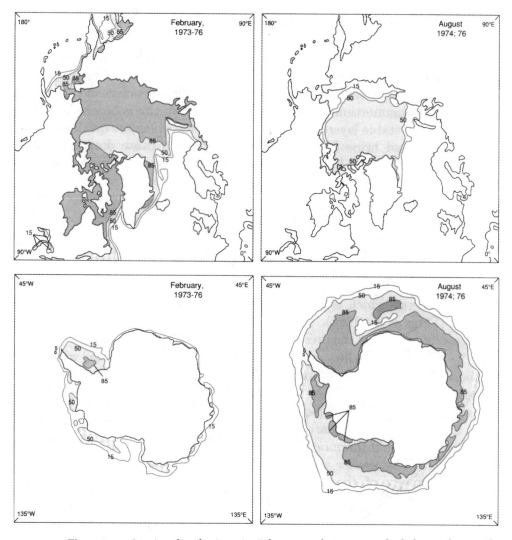

Figure 3.14 Sea ice distributions in February and August in both hemispheres. The contours displayed are 15, 50 and 85% coverage (Trenberth, 1992, Fig. 4.12).

Antarctic pack ice. The annual cycle has an amplitude of some fifteen million square kilometres from a maximum extent of about eighteen million square kilometres and around three million square kilometres in late summer (see Fig. 3.14). From year to year the extent of the ice cover can fluctuate by several million square kilometres (see Section 6.2).

Accurate measurements of the global extent of snow and ice are only available for the satellite era from the early 1970s (see Figs. 3.13 and 6.8). Apart from a puzzling tendency for changes in sea ice cover to show different trends in the northern and southern hemisphere, and a more predictable correlation between winter temperatures and snow cover in the northern hemisphere, there is little evidence of these changes exerting a controlling influence on recent changes in the climate.

Snow and ice cover changes have a substantial short-term impact on the energy balance during the winter half of the year but, in general, the longer term variability of the global climate appears to overwhelm the temporary influence of these variations. In effect, the climatic inertia of a metre or so of snow or pack ice is small compared with the temperature fluctuations of the top hundred metres of the oceans. However, we do not know how other components in the climatic system will respond to changes in snow and ice cover. For instance, the shift in the extent of Antarctic pack ice could produce parallel shifts in the storm tracks at lower latitudes and hence alter the cloud cover. Depending on the form of these additional responses, the net effect could be to either reinforce the changes in the pack ice extent or largely cancel them out. More important may be how these changes couple with shifts in the behaviour of the oceans.

3.7   ATMOSPHERE–OCEAN INTERACTIONS

Moving on from how fluctuations in atmospheric circulation, together with associated variations in cloud cover, and extent of snow and ice, could exert longer term influences on the climate, the next step is how the atmosphere affects the oceans and vice versa. Because of the much greater heat capacity of the oceans, how heat is taken up, stored and released by them is bound to have a bigger impact on longer term climate change. Large-scale temperature anomalies can last much longer than the more fleeting change in snow cover and pack ice. For this reason, the processes which control the surface temperature of the oceans hold the key to many aspects of climate variability and climate change in timescales from a few years to centuries.

The changes in the oceans, cannot, however, be considered in isolation. They are linked with the effects that have been discussed earlier. Long-term fluctuations in cloudiness affect how much energy is absorbed by the oceans, especially in the tropics. Changes in the extent of pack ice may influence the rate at which cold dense water descends into the depths in polar regions. Sustained changes in precipitation and rates of evaporation at high latitudes may have similar consequences. Because these changes may take decades or centuries before influencing the temperature of upwelling of cold water at lower latitudes, they have the capacity to establish longer term fluctuations. But most important of all is that changes in atmospheric conditions can lead to changes in the oceans' surface, which in turn can alter the weather patterns. These atmosphere–ocean feedback mechanisms have the potential to set up oscillatory behaviour and so produce periodicities or quasi-periodicities in the weather, and are central to much of the observed climate variability over the last 100 years or so.

The most celebrated of these feedback mechanisms occurs in the Pacific Ocean. 'When pressure is high in the Pacific Ocean, it tends to be low in the

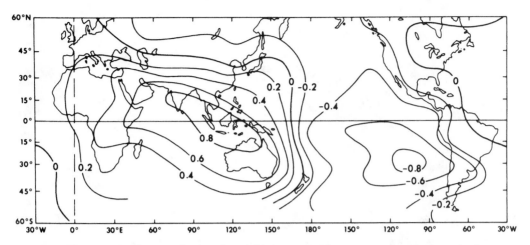

Figure 3.15 The correlation of monthly mean surface pressure with that of Jakarta. The correlation is large and negative in the South Pacific and large and positive over India, Indonesia and Australia. This pattern defines the Southern Oscillation (Burroughs, 1994, Fig. 3.6).

Indian Ocean from Africa to Australia.' This is how Sir Gilbert Walker described what he called the Southern Oscillation (SO) in the 1920s. He defined the SO in terms of differences in pressure observations at Santiago, Honolulu and Manila, and those at Jakarta, Darwin and Cairo. Subsequently, in the 1950s the Dutch meteorologist Berlage updated the index in terms of the overall tropical pressure field. Taking Jakarta as his reference station, he produced a map of the correlation of annual pressure anomalies (see Fig. 3.15) which showed that the value at Easter Island had a surprisingly large value of −0.8 (see Section 7.3). This analysis shows the SO is a barometric record of the exchange of atmospheric mass along the complete circumference of the globe in tropical latitudes. This phenomenon is now defined in terms of the difference in surface pressure anomalies between Tahiti and Darwin (the Southern Oscillation Index [SOI]).

The name El Niño comes from the fact that a warm current flows southwards along the coasts of Ecuador and Peru in January, February and March; the current means an end to the local fishing season and its onset around Christmas means that it was traditionally associated with the Nativity (El Niño is Spanish for the Christ Child). In some years, the temperatures are exceptionally high and persist for longer, curtailing the subsequent normal cold upwelling seasons. Since the upwelling cold waters are rich in nutrients, their failure to appear is disastrous for both the local fishing industry and the seabird population. The term El Niño has come to be associated with an extensive warming of the upper ocean in the tropical eastern Pacific lasting three or more seasons. Because these warm episodes are intimately associated with the Southern Oscillation, the overall behaviour is generally described as the El Niño Southern Oscillation (ENSO).

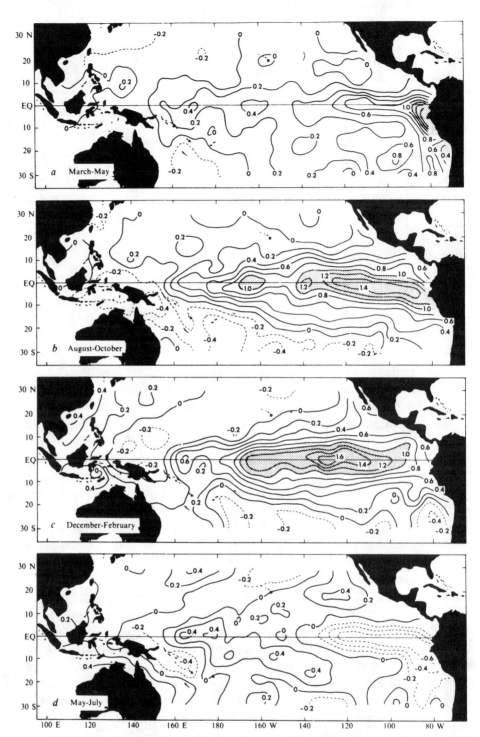

Figure 3.15 Sea-surface temperature anomalies (°C) during a typical El Niño event obtained by averaging the events between 1950 and 1973. The progression shows **(a)** March, April and May after the onset of the warm episode; **(b)** the following August, September and October; **(c)** the following December, January and February; and **(d)** the declining phase of May, June and July more than a year after the onset (Burroughs, 1994, Fig, 5.7).

The oscillatory nature of ENSO means that between the warm episodes there are times when SSTs in the eastern tropical Pacific fall below normal. Although these cold episodes tend to involve less extreme temperature anomalies, they are an important part of the whole ENSO process. Because they are a natural complement of El Niño warm episodes, they have become known as La Niña (Spanish for little girl).

In normal circumstances a warm episode follows a rather well-defined pattern (see Fig. 3.16). In the ocean the onset is marked by above-average surface temperatures off the coast of South America in March to May. This area of abnormally warm water then spreads westwards across the Pacific. By late summer it covers a huge narrow tongue stretching from South America to New Guinea. By the end of the year the centre of the elongated region of warm water has receded to around 130° W on the equator and temperatures are returning towards normal along the coast of South America. Six months later the warm water has largely dissipated and in the eastern Pacific it has fallen below the climatological normal.

In parallel with these changes in SST, large atmospheric shifts are in train. The surface pressure and wind and rainfall records reveal that, starting in the October and November before the onset of the El Niño, the pressure over Darwin, Australia, increases and the tradewinds west of the dateline weaken. At the same time, the rainfall over Indonesia starts to decrease, but near the dateline it increases. In addition, the narrow band of rising air, cloudiness and high rainfall known as the Intertropical Convergence Zone (ITCZ), which girdles the globe, shifts position. Normally, it migrates seasonally between 10° N in August and September and 3° N in February and March. As a precursor to a warm episode it shifts further south in the eastern Pacific, to be close to or even south of the equator during the early months of El Niño years.

As the area of anomalous SST spreads westwards, a region of exceptionally high rainfall associated with the shift of the ITCZ accompanies it (see Fig. 3.17). During the mature phase of a warm episode, most of the tropical Pacific is not only covered with unusually warm surface water but has also exceptionally weak trade winds associated with the southward displacement of the ITCZ. Moreover, the heat transfer from the ocean means that the entire tropical troposphere in the region is exceptionally warm. This maintains the abnormal rainfall until the temperature of the surface waters cool to more normal values. With this return to normality the atmospheric patterns lapse back into a more standard form.

In parallel with these sea-surface and atmospheric changes, important developments occur beneath the surface. The tropical Pacific can be regarded as a thin layer roughly 100 m thick, of warm light water sitting on top of a much deeper layer of colder denser water. The interface between these two layers is known as the *thermocline*. High SSTs correlate with a deep thermocline and vice versa. As the warm episode develops, the easterly trade winds that normally drive the currents in the equatorial Pacific

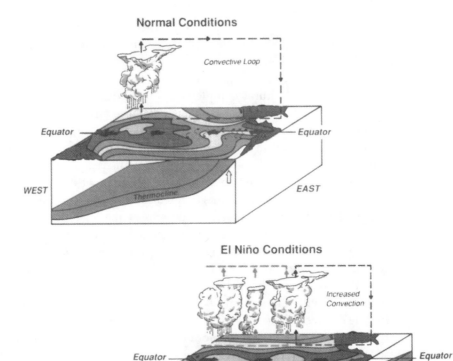

Figure 3.17 Schematic illustration of the changes that take place between normal conditions and those associated with an El Niño event (see Fig. 3.15). The important features are the thermocline becomes less tilted as the sea-surface temperature increases in the eastern Pacific and the increased convection in the central and eastern Pacific (IPCC, 1995, Fig. 4.7).

become exceptionally weak. The sea level in the west Pacific falls and the depth of the thermocline is reduced. Intense eastwards currents between the equator and $10°$ N carry warm waters away from the west Pacific. Along the western coast of the Americas there is an increase in sea level that propagates poleward in both hemispheres. This motion, which may be associated with cyclone pairs in the atmosphere north and south of the equator that reinforce the early flow, creates an eastward propagation motion or wave. This is called 'Kelvin wave' (named after Lord Kelvin in recognition of his fundamental work in wave dynamics), and is confined to close to the equator as its latitudinal extent is restricted by the Coriolis force.

These changes in the oceans contain two important pieces of information. First, the observed movement in the oceans is a consequence of the alteration of the winds which normally drive the currents away from South

America. This standard pattern produces lower sea levels in the east than in the west. It also means that cold water is drawn from higher latitudes and also from greater depths. As the winds weaken so does the current. Sea levels rise and warm water spreads back to cover the cold water. This leads to a second counter-intuitive observation. The development of the SST anomaly appears to reflect a westward movement of warm water which is not the case. The anomaly first appears off the coast of Peru, reflecting the fact that a small reduction in the overall movement westwards can lead to the cold Humboldt current being capped by warmer waters. But the much more extensive region of abnormally warm water across most of the equatorial Pacific only develops after a sustained movement of warm water from the west Pacific. So, although the anomaly appears to move westwards, it is the counter-movement of water that is causing the observed effects. This underlines the central fact about the ENSO that only by considering the combined atmosphere–ocean interactions is it possible to understand the overall behaviour of this phenomenon.

During La Niña events the overall pattern is effectively reversed. Colder than normal SSTs extend across much of the central and eastern Pacific, with the result that atmospheric pressure rises. At the same time pressure over Indonesia and northern Australia falls, and the pressure difference between Tahiti and Darwin (the SOI) becomes positive. This shift strengthens the easterly trade winds and suppresses cloudiness and rainfall over the eastern and central Pacific, while it is enhanced over Indonesia, Malaysia and northern Australia. The same reversal occurs in the ocean. The strengthening trade winds reduce the depth of the thermocline in the east and increase it in the west, while the sea level falls in the east and rises in the west. So the overall impact of the ENSO in the tropical Pacific is for the coupled ocean–atmosphere system to slosh back and forth over the years with huge implications for the climate.

While the nature of the ENSO can be described principally in terms of changes in the tropical Pacific, its impact spreads far and wide. The effects are most noticeable elsewhere in the tropics, as can be seen in the patterns of rainfall (see Fig. 3.18). The distribution of rainfall shifts all around the globe as the changes in pressure associated with the Southern Oscillation alter not only the position of the ITCZ in the Pacific but also less well studied longitudinal atmospheric circulations. During warm episodes when the region of heavy rainfall over Indonesia moves eastwards to the central Pacific, there is a smaller but significant move of the heaviest rainfall over the Amazon to the west of the Andes. More important is that the region of ascending air over Africa is replaced by a descending motion. This partially explains how the prolonged drought in sub-Saharan Africa since the late 1960s has been related to more frequent El Niño events in the Pacific.

The ENSO is also involved in another great mystery of tropical meteorology – the Indian monsoon. Ever since Edmund Halley proposed in a paper in the Philosophical Transactions of the Royal Society in 1686 the broad

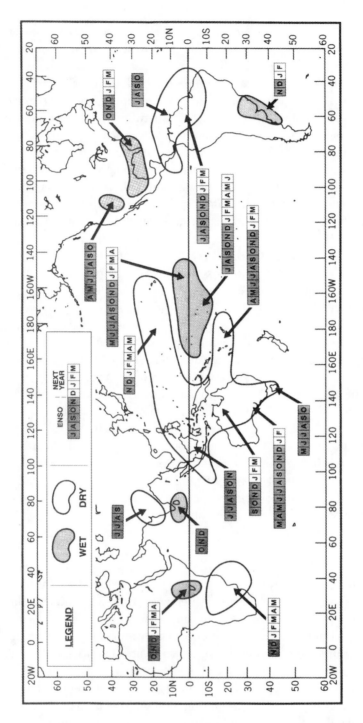

Figure 3.18 Schematic diagram of the areas and times of the year with a consistent ENSO precipitation signal (IPCC, 1990, Fig. 7.21(a)).

mechanism for the monsoons, there has been speculation as to what caused its fluctuation from year to year. It has now become clear that the ENSO is a factor in whether the summer monsoon brings abundant rains or drought to the Indian subcontinent. When the tropical Pacific is warmer than usual (the El Niño years) rainfall is often below normal in India and vice versa (see Fig. 3.19). Events in the 1970s and 1980s appeared to reinforce the impression that had built up since the 1920s that this link was a useful predictor for monsoon rainfall. More recently, notably during the huge El Niño event of 1997, the relationship has proved less reliable. Observations during 1997 and 1998 suggest that one explanation is something which some meteorologists have long suspected, namely that SST patterns in the Indian Ocean are another important factor, and this ocean may also be capable of quasi-oscillatory behaviour. This additional input may also be crucial in understanding what controls longer-term rainfall patterns over southern Africa.

The scale of these changes in the tropics and the fact that in recent decades the global warming has been closely linked to ENSO behaviour suggests there should be parallel disturbances at higher latitudes. These are not immediately obvious. While some features of year-to-year variability in the climate of mid-latitudes can be linked with events in the equatorial Pacific, they constitute only a small part of the story. Indeed, in the North Atlantic in particular, there is another less well understood 'oscillation', which may be a further important aspect of atmosphere–ocean interactions. Since the eighteenth century it has been known that when winters are unusually warm in western Greenland, they are severe in northern Europe and vice versa. This see-saw behaviour was quantified by Sir Gilbert Walker in the 1920s in terms of pressure differences between Iceland and southern Europe and he defined it as the North Atlantic Oscillation (NAO).

The NAO shifts between a deep depression near Iceland and high pressure around the Azores, which produces strong westerly winds, and the reverse pattern with much weaker circulation. The strong westerly pattern pushes mild air across Europe and into Russia, while pulling cold air southwards over western Greenland. The strong westerly flow also tends to bring mild winters to much of North America. One significant climatic effect is the reduction of snow cover, not only during the winter, but well into the spring. The reverse meandering pattern often features a blocking anticyclone over Iceland or Scandinavia which pulls arctic air down into Europe, with mild air being funnelled up towards Greenland (see Fig. 3.8). This produces much more extensive continental snow cover, which reinforces the cold weather in Scandinavia and eastern Europe, and often means that it extends well into spring as long as the abnormal snow remains in place (see Section 3.6).

Since 1870, the NAO has fluctuated appreciably on timescales from several years to a few decades (see Fig. 3.20). It assumed a strong westerly form between 1900 and 1915, in the 1920s and, most notably, from 1988 to

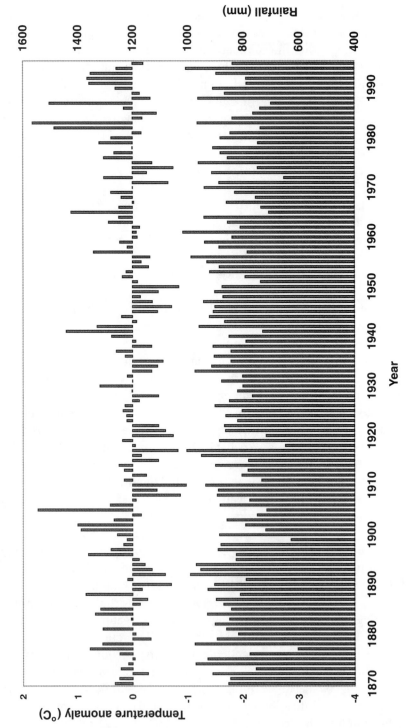

Figure 3.19 Variations in the all-India summer monsoon between 1871 and 1994, together with an indication of the principal El Niño events. This shows that years of low rainfall tend to be linked with the onset of strong El Niño events.

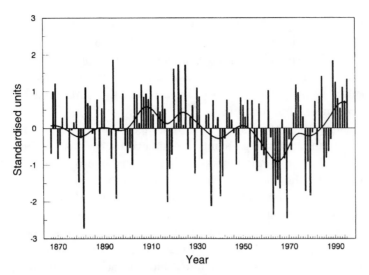

Figure 3.20 The North Atlantic Oscillation as measured by the standardised difference of the December to February atmospheric pressure between Punta Delgada, Azores and Stykkisholmur, Iceland, together with a smoothed curve to show fluctuations longer than around ten years (IPCC, 1995, Fig. 3.18).

1995. Conversely, it took on a sluggish meandering form in the 1940s and during the 1960s bringing frequent severe winters to Europe but exceptionally mild weather in Greenland. So it appears to be a more persistent phenomenon than interannual fluctuations in snow and ice cover. How these variations might play a part of an organized atmospheric–ocean interaction is the subject of intense research at present.

Changes in sea–ice cover in both the Labrador and Greenland Seas as well as over the Arctic appear to be well correlated with the NAO, and the relationship between the sea-level pressure and ice anomaly fields suggests that atmospheric circulation patterns force the sea ice variations. Feedbacks or other influences of winter ice anomalies on the atmosphere have been more difficult to detect. But, the frequency of depressions appears to have increased and atmospheric pressure decreased where ice margins have retreated, although these changes differ from those directly associated with the NAO. If, however, it can be demonstrated that a period of one phase of the oscillation produces the right combination of patterns of sea surface temperatures and deep water production, eventually to switch it into the opposite phase, then this may provide insight into more dramatic changes in ocean circulation which can occur in the North Atlantic (see Section 3.8).

Recent research clearly shows that the North Atlantic responds to changes in the overlying atmosphere. The leading mode of sea surface temperature variability over the winter consists of a tripole: a cold subpolar region, warmth in the middle latitudes, and a cold region between the equator and 30° N. This ties in with shifts in the surface fluxes associated with the NAO pattern. The NAO index appears to lead the sea surface temperatures indicating that they are responding to atmospheric forcing on monthly timescales. Climate model studies suggest, however, that the longer term tripole sea surface temperature pattern drives the NAO, but provide less insight into the extent to which the NAO links back into sea surface temperature patterns.

Whatever the impact of the NAO on longer term developments, it remains central to understanding recent climatic events, because of the influence it exerts on average temperatures in the northern hemisphere. Of all seasons, winters show the greatest variance, and so annual temperatures tend to be heavily influenced by whether the winter was very mild or very cold. When the NAO is in its strong westerly phase, its benign impact over much of northern Eurasia and North America outweighs the cooling around Greenland, and this shows up in the annual figures. So, a significant part of the global warming since the mid-1980s has been associated with the very mild winters in the northern hemisphere. Indeed, since 1935 the NAO on its own can explain nearly a third of the variance in winter temperatures for the latitudes 20° to 90° N.

The hemispheric nature of the warming patterns also show up in the satellite data of snow cover (see Fig. 3.13). The monthly variations in snow cover across northern Eurasia and North America show a significant correlation. This is not surprising given how strong westerly circulation will be associated with above-normal winter temperatures and below-normal snow cover. But the fact that the snow cover responds on a hemispheric scale reinforces the climatic forcing of these fluctuations. So understanding more about how the NAO acts as a major natural factor in climate variability is central to explaining the causes of global warming in the twentieth century.

While the ENSO and the NAO have attracted a great deal of attention in recent years, there is no reason why atmosphere–ocean interactions in other parts of the world should not produce similar longer term climatic fluctuations. Indeed, the upsurge in interest in these interactions has led climatologists to look much more closely for evidence of such connections elsewhere. The product of this work has been a growing list of 'oscillations' from the Arctic to the Antarctic. The real test of these new teleconnections will be whether they offer new insights into the workings of the climate and better still if they are able to provide useful forecasts months and years ahead.

Perhaps the potentially most interesting of these oscillations is in the Pacific Ocean, which covers nearly a third of the Earth's surface. This huge area, combined with its roughly symmetrical form (both latitudinal and longitudinal) make it the most obvious region for longer term coupled interdecadal variations of the atmosphere and ocean. While shorter term fluctuations in the two- to ten-year range are dominated by the ENSO variation, interdecadal changes have the largest amplitude in the north Pacific rather than in the eastern tropical Pacific, together with a coherent pattern of surface temperature variability in the Southern Hemisphere with cold (warm) anomalies in the region of New Zealand and warm (cold) anomalies in the south eastern tropical Pacific. Known as the Interdecadal Pacific Oscillation (IPO), these changes exhibit an approximately twenty-year periodicity with the possibility of an underlying longer fifty- to sixty-year variation. Recent work suggests that these longer period variations can also modulate ENSO teleconnections.

3.8 THE GREAT OCEAN CONVEYOR

So far the consideration of the part played by the oceans has concentrated largely on their surface properties, horizontal motion and the interaction with the atmosphere. This has effectively restricted the analysis to the mixed layer of the oceans generated by the action of the winds. From top to bottom of this mixed layer there is little temperature difference. As noted in Section 3.7, beneath it is the thermocline – a narrow zone over which there is a rapid drop in temperature. The thickness of the mixed layer, and hence the depth of the thermocline, depends on the wind speed, thermal mixing where the surface waters are heated by the sun or altered by the passage of warmer or colder air, and by the advection of warmer or colder water or the upwelling of cold water. However, this involves only a small part of the oceans. Beneath the seething surface layer is a much more gradual but equally important set of motions.

The process driving the deep waters of the oceans is *thermohaline circulation*. This results from changes in sea water density arising from variations in temperature and salinity (see Box 3.1). Where the water becomes denser than the deeper layers it can sink to great depths. The temperature depends on where the surface waters come from and how much heat the oceans either pick up or release to the atmosphere in both sensible heat and evaporative loss. The salinity of a given body of water depends on the balance between losses through evaporation as opposed to gains from either rainfall, or freshwater run-off from rivers and melting of the ice sheets of Antarctica and Greenland plus the pack-ice of the polar oceans. In practice there are few regions where sinking waters have a major impact. *Deep waters*, defined as water that sinks to middle levels of the major oceans are formed only around the northern fringes of the Atlantic Ocean. *Bottom waters*, which constitute a colder denser layer below the deep waters, are formed only in limited regions near the coast of Antarctica in the Weddell and Ross Seas.

Thermohaline circulation drives a worldwide pattern known as the Great Ocean Conveyor (GOC) (Fig. 3.21). This model, which was developed by Wallace Broecker, at the Lamont Doherty Laboratory, has had an immense impact on recent thinking about the nature of climate change. Although it

BOX 3.1 THERMOHALINE CIRCULATION

The density (ρ) of sea water is not a simple function of temperature and salinity. So the easiest way to visualise how changes in these parameters control thermohaline circulation is to consider the actual values of the density set out in Table B3.1. These show that while freshwater has a maximum density at 4 °C, at normal levels of salinity (32.5–37.5‰) the density increases with declining temperature right down to the freezing point around −2 °C. The changes in density with salinity are simpler as, the saltier the water, the more dense it becomes.

TABLE B3.1 DENSITY OF WATER OF DIFFERENT SALINITIES AT DIFFERENT TEMPERATURES (DENSITY UNITS ARE SIGMA VALUES,[a] AT ATMOSPHERIC PRESSURE)

Temp. (°C)	Salinity (‰)							
	0	10	20	30	32.5	35	37.5	40
30	−4.3	3.1	10.6	18.0	19.9	21.7	23.6	25.5
20	−1.6	5.8	13.3	21.0	23.0	24.7	26.6	28.5
10	−0.3	7.6	15.2	23.1	25.2	26.9	28.8	30.8
5	0	8.0	15.8	23.7	25.7	27.6	29.6	31.6
0	−0.1	8.0	16.1	24.1	26.1	28.1	30.2	32.2
−2	−0.3	7.9	16.0	24.2	26.1	28.2	30.3	32.3

[a] The sigma value (σ) is a way of presenting the density (ρ) of sea water relative to that of distilled water at 4 °C, where $\rho = 1.000$:

$$\sigma = 1000(\rho - 1)$$

Thus the densities given in Table B3.1 range from 0.9957 (at 30 °C and zero salinity) to 1.0323 (at −2 °C and 40‰ salinity).

Several basic observations can be made about these figures. First, where freshwater enters the oceans (e.g. rivers, melting ice and rainfall) it can float on the top of sea water preventing deep circulation. At the same time any surface warming due to either solar heating or advection of warmer water from lower latitudes will tend to form a stable surface layer. The stability of this surface layer, and hence the depth of the thermocline will be controlled by the amount of surface mixing due to winds. But deep mixing depends on the formation of high salinity water either by the process of freezing as salt is shed into the surrounding cold water or in areas of high evaporation. These processes (e.g. the creation of North Atlantic Deep Water, and Antarctic Bottom Water, and the flow in and out of the Mediterranean, whose dense salty water spills out into the deep Atlantic and is replaced by less dense Atlantic surface water) are the principal mechanisms driving larger scale thermohaline circulation of the oceans.

For the rest, the surface motions of the oceans reflect the effects of the winds and the differing salinity and temperature of various bodies of water that mix over time. This is a difficult process to represent in computer models (see Section 9.1) because much of this mixing takes place in the form of relatively small eddies. So the density of sea water must be calculated as its properties change. This involves the calculation of *isopycnic surfaces* (defined by points of equal density). For instance sea water with a salinity of 32.5‰ and a temperature of 0 °C has the same density as that with values of 35‰ and 14.5 °C, or 37.5‰ and 22 °C (see Table B3.1).

may turn out to be an oversimplified representation, and that the real world is more complicated, it provides an excellent starting point for considering the climatic implications of thermohaline circulation. In particular, it ties in neatly with the surface ocean currents which transport so much energy polewards (see Section 2.1.4). So any changes in the scale of this process will have major consequences for the climate.

There is evidence that the GOC can exist in a number of very different states and that this could be a fundamental factor in controlling climatic change. This is of particular relevance to the North Atlantic where the current circulation carries the majority of heat to the Arctic (see Fig. 2.6). This

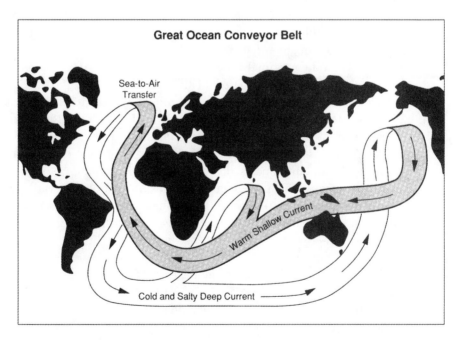

Figure 3.21 The Great Ocean Conveyor Belt – a schematic diagram depicting global thermohaline circulation (Trenberth, 1992, Fig. 17.12).

warm water gives up its heat to the arctic air through evaporation. Then its low temperature and high salinity enables it to sink and form deep water which flows all the way to Antarctica. Here it is warmer and less dense than the locally frigid surface waters and so rises to become part of a strong vertical circulation process. Descending cold water from around Antarctica flows northwards into the Pacific and Indian Oceans where there is no descending cold water. In the Atlantic this countercurrent is swallowed up by the much stronger southward flow.

Knowledge of the precise form of this deep water flow is still the subject of experimental analysis. Recent oceanographic research is beginning to show how the NAO might be part of a quasi-cyclic fluctuation in the North Atlantic. The prolonged period of a positive NAO in the winters from 1988 to 1995 provided interesting clues (see Fig. 3.20). This change has pushed warmer, fresher Norwegian Atlantic water farther north into the Fram Strait and Barents Sea. Most recently, a warmer, fresher surface layer, thinning sea ice and a record extent of open water have affected the western Arctic. At the same time, the prevailing north westerly airflow in the Davis Strait has created an exceptionally deep body of cold water in the Labrador Sea. Measurements show that this water has spread out across the North Atlantic at depths of 1000–2000 m during the period 1990–95. At the same time the creation of deep water in the Greenland Sea, north of Iceland, was largely cut off as was the deep water circulation in the Sargasso Sea. In stark contrast, during the 1960s, when the NAO was in a strongly negative phase, there was deep convection in the Greenland and Sargasso Seas and in the Labrador Sea convection was tightly capped with a build-up of relatively fresh water at the surface. This capacity to switch back and forth between two such different states has major implications for climatic fluctuations on the decadal and centennial timescales.

Elsewhere, the representation of the easterly flow of the GOC beneath the Southern Ocean in Fig. 3.21 is probably oversimplified, not least because bottom water flowing northwards is channelled through a number of relatively narrow gaps in the pronounced ridge on the sea floor which circles Antarctica. Nevertheless, the general form of the circulation brings out three important aspects of the processes influencing climate change. First, the oceans transport huge quantities of energy polewards. Second, at high latitudes in the northern hemisphere the Atlantic represents some 60% of the oceanic energy transfer (see Fig. 2.6). Third, any changes in the amount of energy carried by the oceans, especially the North Atlantic, can have a major impact on the climate of the northern hemisphere. Furthermore, there is plenty of evidence to show that the circulation has made sudden shifts in the past (see Section 4.5), so the consequences for the future could be dramatic (see Section 8.3). Hence, a better understanding of how the deep oceans circulate and what factors can lead to changes in these movements is crucial in interpreting past shifts in the climate and how it may develop in the future.

3.9   SUMMARY

The essential feature of the various elements of the Earth's climate is how they are all connected to one another. So, it is not realistic to isolate one element and consider what changing it might do to the climate. Although this may provide some interesting insights, failure to estimate how its knock-on effects will affect the rest of the system is a recipe for failure. Only by considering the Earth's climate as a whole is it possible to form a sensible appreciation of how it has changed in the past and how it may alter in the future.

The network of links in the climate is the key to the balanced view. In building up this picture we have considered how each element affects:

a) the amount of solar energy that is absorbed, noting in particular the importance of clouds, snow and ice and land surface effects, including the role of the hydrological cycle and the biosphere;

b) how energy is transported across the face of the Earth by both the atmosphere and the oceans underlining the importance of the atmosphere to assume long-lasting circulation patterns and the oceans to alter how they move and store energy; and

c) how the amount of energy re-radiated to space is controlled by various elements, again placing particular emphasis on the role of clouds.

The timescale of the response to all these processes to change is also central to how they contribute to fluctuations in the climate. For the most part, the atmosphere reacts more rapidly to various stimuli, but is capable of getting stuck in long-term states which are capable of producing significantly different climatic conditions over large parts of the globe. But, in every aspect of atmospheric variability, the role of clouds is crucial. By comparison, oceans effectively respond more slowly but are capable of producing much greater changes. The speed of the response of land surface conditions, and snow cover, tends to fall in between. These differing time constants mean that the various climatic elements will not react in a uniform way to change. Instead there will be lags and leads between them which are capable of producing erratic and sometimes contradictory responses throughout the system.

How the various elements of the climate can combine to produce change, has to be put in context with what is known about past changes in the climate. This involves reviewing the evidence, consequences and measurement of variability and change to establish what can reasonably be expected to happen. This is a journey through vast stretches of time, viewing an extraordinary range of clues which provide some indication of the complexities which must be addressed. But, none of these alters the fact that what we are really concerned about is how the fluctuating behaviour of the atmosphere and the oceans combine to produce such a variety of climatic regimes.

QUESTIONS

1 Explain the radiative arguments for the statement that low clouds cool the Earth's climate whereas high clouds warm it.

2 With the aid of a globe explain why the jet stream has a wave-like pattern rather than being circular as it circles the Earth.

3 It can be argued that reductions in the extent of sea ice will lead to a strong positive feedback as the exposed sea water will be able to absorb more solar energy and so lead to further warming. What other processes might lead to a negative feedback which results in much of the sea ice effect being cancelled out?

4 Using the albedo figures in Table 2.1 make an estimate of the difference between the amount of energy absorbed by a unit area of the Sahara desert and the same area covered by savannah vegetation. Is this difference significant and, if so, is it a realistic assessment of the climatic impact of the expansion of desert in this part of the world?

FURTHER READING

A complete reference list is available at the end of the book but the following is a selection of the best books or articles to follow up particular topics within this chapter. Full details of each reference are to be found in the Bibliography.

Barry and Chorley (1992). An excellent textbook for discovering the basics of how the global climate functions, but best read in conjunction with other texts when it comes to climatic change.

Bigg (1996). Particularly valuable in providing analysis of the growing understanding of how the atmosphere and oceans combine to govern many aspects of the climate.

Bryant (1997). A new textbook which takes a slightly different approach to the question of the climate change in exploring some of the difficult questions of whether we adequately understand the causes of change and can model their consequences. This inquisitive approach, combined with a clear and direct style, makes it a valuable source for getting to grips with a number of difficult issues.

Open University Oceanography Series (1989). The ocean circulation volume, in particular, provides a clear and concise introduction to the basic aspects of ocean dynamics.

Philander (1990). A thorough and penetrating survey into the nature and causes of large-scale climatic fluctuations in the tropical Pacific and their influences on global climate.

CHAPTER FOUR

EVIDENCE OF CLIMATE CHANGE

Time, which antiquates antiquities, and hath an art to make
dust of all things, hath yet spared these minor monuments.

Sir Thomas Browne, 1605–1682

The first step in understanding the nature of fluctuations in the climate is to
examine the nature of the evidence of change. Much of what has been used
to support theories of climate change is circumstantial and fragmentary: in
effect only a few pieces of a jigsaw. So differing interpretations can be put on
what actually could have occurred and what the causes might have been for
the proposed changes. These will be continually evaluated as more evidence
becomes available or improved measurement techniques enable the existing
evidence to be reappraised in a more critical light. For example, the holding
of Frost Fairs on the frozen River Thames in London during the seventeenth
and eighteenth centuries have long been seen as confirmation of the winters
being much colder then than of late. But, how much can we read into a few
extremely cold seasons? Moreover, how do we take account of the fact that
the old London Bridge, which was removed in 1831, acted as a weir to slow
down the flow of the river? Combined with the absence of embankments,
which meant the river was much wider, and the lack of waste heat from
industrial plants, ensured that the river froze much more readily in cold
weather. Only with the careful examination of instrumental records (see
Section 6.1) did it become clear that winters were on average 1 °C colder:
a smaller difference than might be inferred by relying on the records of Frost
Fairs alone.

Similarly, evidence of ancient civilisations thriving in areas which are
now desert have to be examined closely to ensure that other factors, which

could have contributed to their collapse, have been adequately analysed. In the same way, the evidence that mountain glaciers were more extensive in the past shows clearly that periods of colder, wetter weather led to more snowfall at high levels. But, how precisely this happened and whether successive expansions and contractions of the glaciers erased all evidence of previous surges makes it difficult to construct a clear record of climate change over time. The same applies to the evidence of the formation of huge ice sheets over large areas of the northern hemisphere which shows beyond a shadow of doubt that there have been a series of ice ages during the last million years.

Ideally, evidence should be based on an accurate measure of a specific meteorological parameter at a given spot at a known time. With enough measurements it is possible to build up a picture of how the climate has changed with time. In practice, measurements rarely achieve this goal. Even with modern instrumental observations there are gaps, and these grow rapidly as we go back in time. But for almost all of the Earth's history we must rely on indirect (*proxy*) measurements, such as tree rings, ice cores, and lake and ocean sediments. But even with the exploitation of a wide variety of sources, there are vast gaps in the record.

Exploring the evidence of climatic variations over the complete history of the Earth covers processes that have taken an immense time to occur (e.g. continental drift). In terms of current concerns about future changes in the climate these changes seem immeasurably slow and hence of little relevance to contemporary issues. This view is short-sighted. An understanding of longer term changes not only sets current events in context, but also identifies the importance of different components of the global climate (e.g. changes in ocean circulation). So the more we know about what has happened in the past, and why they occurred, the easier it may be to appreciate the questions that need to be addressed today.

This catholic approach means we need to be clear what we are trying to do in this chapter. The objectives are to identify the principal evidence of climate change and to expose the limitations in our knowledge of what changes have occurred. This approach will prepare the ground for considering the consequences and causes of climate change, and whether we can take a view on how it may change in the future.

4.1   PEERING INTO THE ABYSS OF TIME

Much of the early realisation that the Earth's climate had undergone vast changes in the past emerged with the development of the science of geology. As geologists started to build up models of how the stratigraphy of the Earth had been laid down over the vastness of time, it became clear that huge changes had occurred in the climate. More recently, it was realised that the distribution of the continents had changed over time.

So the interpretation of the climatic clues locked up in the rocks has been a painstaking business. This involved building up a picture of the conditions prevailing at the time the sediments were deposited and whereabout on the globe these processes took place.

The geological evidence of fossilised plant and animal remains, organic material, pollen and the shells of creatures contained in sedimentary rocks provide many clues. Working on the basic principle that some idea of past climate can be ascertained by comparing these rocks with similar sediments being deposited today. In early parts of the record, geologists draw on whatever sediments are available to build up a consistent picture. But, as we approach the current distribution of the continents the analysis of ocean sediments can be more discriminating because the topography and oceanic conditions when they were laid down are better known.

In particular, the remains of various forms of phytoplankton are of great value. These occur widely in large numbers in ocean sediments. Because the different forms of plankton are readily identifiable and are known to have lived in certain conditions (i.e. warm or cold waters, and near the surface or great depth) their distribution and physical properties (i.e. isotopic composition – see Section 6.4.3) provides detailed information about climatic conditions during their lifetimes. For example, a few groups of oceanic plankton living near the surface of the ocean can be used to subdivide the last sixty-five million years (My) in no fewer than fifty to sixty zones.

The nature of the rocks can provide a variety of climatic information. The remains of dried-out lakes and seas (*evaporites*) are signs of prolonged aridity. Large deposits of *carbonate rocks* indicate periods when the carbon dioxide in the atmosphere was drawn down. Glacial tills, which contain a poorly sorted mixture of soil and rocks, become consolidated to form *tillites*, while ice will transport boulders from identifiable sources across the landscape and leave them in unlikely places (*erratics*) or carry them out to sea before melting to leave *ice-rafted debris* in ocean sediments. Direct physical indications of the movement of ice in the form of *striations* on rocks provide evidence of the presence and extent of glaciers and ice sheets.

In building up a consistent explanation of past geological developments a number of basic principles apply. First, there is the underlying assumption that the physical processes involved in creating sedimentary layers have neither slowed down or speeded up over time (*uniformitarianism*). So the processes occurring now have been the same throughout geological time. In short – 'The present is the key to the past.' Secondly, successive layers of sediments are superimposed with the youngest at the top and the oldest at the bottom. Moreover, these layers exhibit lateral continuity and were initially horizontal, or nearly so. This means breaks in the record are due to either erosion or vertical movement. Finally, where there are, say, faults or intrusive dikes (due to volcanoes), these features are younger than the sedimentary layers they cut across.

In practice, the interpretation of stratigraphic layers is made much more complicated by the changes that disturb them over time. The combination of crustal movement, which leads to folding, faulting, uplifting and downwarping, together with the continual process of erosion has confused and destroyed much of the record. Estimates of the amount of sediments retained in the stratigraphic records suggest that between 90% and 99% of the material originally laid down has subsequently been eroded. So the resulting record has huge gaps in time and from place to place. This means that for two centuries geologists have been piecing together evidence from the abyss of time. In so doing, their views of such essential matters as the age of the Earth, continental drift, the rise and fall of sea level, and the nature of mass extinctions have changed radically over the years.

The defined geological timescale (Fig. 4.1) has been constructed on these principles. The important thing to concentrate on is the time covered by each period rather than the names which are a wonderful combination of Greek/Latin terms (e.g. *Palaeozoic*, meaning time of old life) and whimsical links with the regions where rocks belonging to given periods were first identified (e.g. *Ordovician* and *Silurian* are named after the ancient British

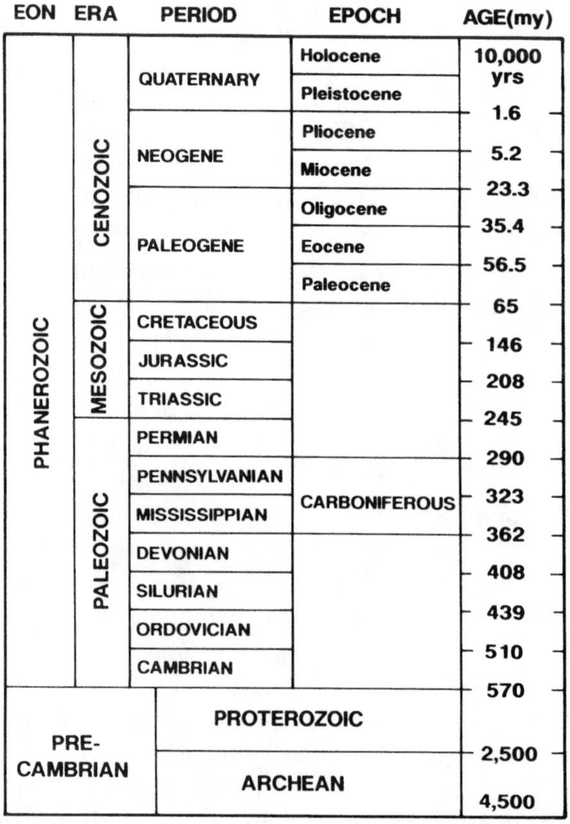

EON	ERA	PERIOD	EPOCH	AGE(my)
PHANEROZOIC	CENOZOIC	QUATERNARY	Holocene	10,000 yrs
			Pleistocene	1.6
		NEOGENE	Pliocene	5.2
			Miocene	23.3
		PALEOGENE	Oligocene	35.4
			Eocene	56.5
			Paleocene	65
	MESOZOIC	CRETACEOUS		146
		JURASSIC		208
		TRIASSIC		245
	PALEOZOIC	PERMIAN		290
		PENNSYLVANIAN	CARBONIFEROUS	323
		MISSISSIPPIAN		362
		DEVONIAN		408
		SILURIAN		439
		ORDOVICIAN		510
		CAMBRIAN		570
PRE-CAMBRIAN		PROTEROZOIC		2,500
		ARCHEAN		4,500

Figure 4.1 The Geological Timescale (Van Andel, 1994, Fig. 2.4).

tribes *Ordovices* and *Silures* which once inhabited the Welsh borders where these distinctive rocks were first identified). Many of the quoted dates are only accurate to a few percent. This means that for the late Cenozoic the boundaries between various periods can be trusted to within 100,000 years or so. This uncertainty increases to around a million years in the Palaeocene and grows to several million years in the early Mesozoic. Nevertheless, for the purposes of maintaining a consistent chronology, dates will be quoted with greater precision to enable adjacent events to be discriminated, as these changes can be resolved with greater precision than the absolute timescales can be defined.

There is little remaining to tell us how the climate changed for most of the first 90% of the Earth's lifetime. We do not know where the oceans and the continents were, or what precisely the constituents of the atmosphere were. The first sedimentary rocks were laid down some 3,700 million years ago (Ma), when it is believed that the climate may have been some 10 °C warmer than now. The first primitive forms of life may have appeared as early as around 3,800 Ma but they provide little evidence of the climate. All we know is that between 2,700 and 1,800 Ma widespread glacial conditions were a feature of the climate for at least part of the time.

In Ontario and Wyoming, for example, there is evidence, in the form of preserved tillites, of three discrete glaciations between 2,500 and 2,200 Ma. Moreover, recent work suggested most of southern Africa was covered by glaciers at this time when that region was close to the Equator. This raises the question that the Earth was largely covered with ice – sometimes referred to as *Snowball Earth* – a condition which might be expected to last forever as it would have reflected most of the Sun's energy back into space. Instead, these glacial rocks are capped with either carbonates or volcanic magma, which can be interpreted as a catastrophic event (e.g. volcanic eruption, overturning of a stagnant ocean or cometary impact – see Chapter 8) which could have pumped enough carbon dioxide back into the atmosphere to melt the planet's icy shell. Whatever the explanation thereafter, it seems the Earth remained warm and free of ice caps and glaciers for around a billion years.

The mists begin to clear around 1,000 Ma. The late Precambrian period experienced a glacial period, which as best we can tell, lasted some 200 My (see Fig. 4.1), sometimes termed the Cryogenian period of the late Precambrian. It appears there were at least two ice ages between 850 and 590 Ma. Analysis of carbonate rocks bracketing these events indicate they may each have involved a number of glacial episodes, some of which could have extended to low latitudes. Indeed the Snowball Earth scenario may have prevailed, although recent computer-modelling studies of conditions at the time suggest that some equatorial oceans remained ice-free. The extent of this ice-free region could have been crucial to both the maintenance of life on Earth, and the course of subsequent evolution of every species. The impact of these huge climatic events is still the subject of great uncertainty. There is, however, no doubt that they encompassed a period of extensive and substantial climatic changes.

Towards the end of the Precambrian period dramatic evolutionary changes began to occur. Around 700 Ma there was a rapid diversification of higher plants and animals with soft bodies. Then, following the last glacial period around 590 Ma, during the Cambrian there was a sudden dramatic acceleration in evolution starting around 565 Ma when all the basic designs for animal life, involving shells and hard skeletons, make their first appearance in the fossil record. This is often termed the *Cambrian explosion*. The scale of these changes and the improving fossil record means that for the purposes of this book our examination of climate change will now be restricted to events from the beginning of the Cambrian. These last 600 My are known as the *Phanerozoic* (the 'Age of Visible Life').

The interpretation of the geological evidence of the Phanerozoic is made easier in recent decades by the parallel development of theories of continental drift. Since measurements of palaeomagnetism of ocean-floor spreading in the 1960s confirmed earlier hypotheses about how the plates, which make up the Earth's crust, move, the history of the migration of the continents has been developed. During the first half of the Phanerozoic the main continent was Gondwanaland, which was made up of Africa, South America, India, Antarctica and Australia (Fig. 4.2). Other continental fragments included parts of what are now North America and Eurasia. Mapping the movement of Gondwanaland has enabled the changes in geological record, which might be attributed to climate change, to be put into context.

During the Cambrian the climate warmed appreciably and remained relatively warm for most of the following 300 My. This era is known as the *Palaeozoic*. There was a relatively brief glacial period around the end of the Ordovician and beginning of the Silurian. The most extensive evidence of this cool period is now to be found in the Sahara (Fig. 4.3), which was close to the South Pole at this time (see Fig. 4.2). During the Carboniferous the temperature dropped, culminating in the long Permo-Carboniferous glaciation from 330 to 250 Ma. This icy epoch may have been the coldest period in the Earth's history. Its later stages coincided with the formation of the supercontinent, Pangaea, when all the Earth's land masses came together, and stretched from the equator to the South Pole, with what are now Antarctica and India at high latitudes forming the centre of glaciation (see Fig. 4.2). Vast areas of this supercontinent were probably covered in ice, although climatic conditions in other parts of the globe may have been rather warm. Also the interior of the unglaciated continents became more arid.

At the same time, the compilation of a coherent explanation of these changes has been complicated by dramatic shifts in the fossil records. Certain cataclysmic events, including possibly asteroid impacts with the Earth, immense volcanic eruptions or climate change have led to mass extinctions of flora and fauna. The five most clearly established examples

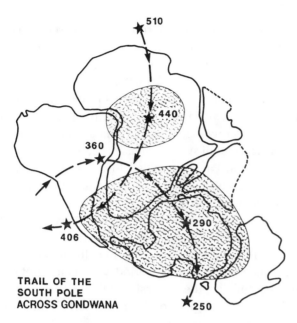

**TRAIL OF THE
SOUTH POLE
ACROSS GONDWANA**

Figure 4.2 Between the late Precambrian and the Permian, the main continent was Gondwanaland, made up of Africa, South America, India, Antarctica and Australia, so its geographical position during this time is crucial to understanding climatic change. The South Pole traversed the continent at least twice (the numbers along its track mark its progress in My), and ice caps formed around the pole (shaded areas), but at other times the polar regions were free of ice (Van Andel, 1994, Fig. 7.3).

Figure 4.3 Ordovician tillites, about 440 My old, from the Saharan desert. The striated boulders are clear indicators of a glacial origin (*Cambridge Encyclopaedia of Earth Sciences*, Fig. 23.23).

of these events (the 'Big Five') were Late Ordovician, 440 Ma; Late Devonian, 365 Ma; Late Permian and Early Triassic, 255 Ma; Late Triassic, 210 Ma; and end-Cretaceous, 65 Ma. The consequences of this process are breathtaking. Over the last 600 My, 99.9% of all species that ever lived have perished. Nevertheless, biological diversity has increased so there are now at least a million species of animal, of which insects make up three-quarters of the total (see Section 5.2). But extracting evidence of climate change from the fossil record, against the background of such immense changes, is a daunting task.

During the Mesozoic era, Pangaea drifted towards lower latitudes and broke up (Fig. 4.4). The standard view is that this was a period of generally warm climate with relatively little difference in temperatures between the poles and the tropics and there were insignificant seasonal variations. This equable period is usually associated with the 'Age of the Dinosaurs'. There is now new evidence, however, of some fluctuations in climate throughout this era, including a cooler period with cold winters at high latitudes around the end of the Jurassic and the beginning of the Cretaceous. This includes examples of ice-rafted debris in Siberia, the

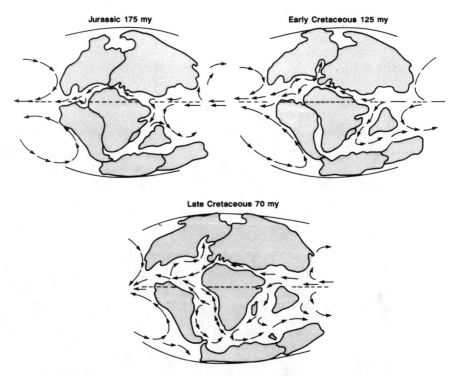

Figure 4.4 During the Mesozoic the supercontinent Pangaea slowly broke up and the surface circulation of the oceans evolved from a simple pattern of a single continent in a single ocean to a more complicated pattern in the new oceans of the Cretaceous. The combination of the open circum-equatorial path and the absence of circum-polar currents during this period produced a more even temperature distribution than today (Van Andel, 1994, Fig. 10.7).

Canadian Arctic, Spitzbergen (Svarlbard) and central Australia, all of which were at high latitudes at this time. There is also evidence of a sharp reduction in sea level around 128 and 126 Ma (see Section 4.3), which suggests glaciation of continental interiors at these latitudes. The greater seasonality of the climate at this time is confirmed by tree-ring studies of fossil wood which show definite annual boundaries, indicating that growth ceased in the winter (see Section 6.4.1).

Following this cooler period, the mid-Cretaceous, around 100 Ma, was probably the warmest period in the Earth's history for which there is reasonable data (Fig. 4.5). Globally, the average temperature is estimated

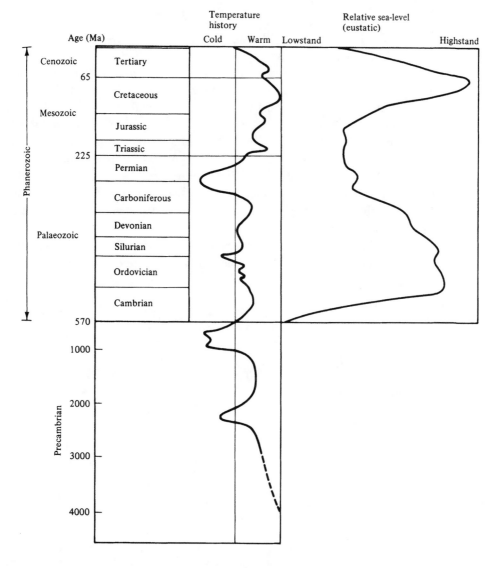

Figure 4.5 A generalised temperature history of the Earth. The changes in sea level for the Phanerozoic are also shown (Brown, Hawkesworth and Wilson, 1992, Fig. 24.7).

to have been 6–12 °C warmer than the present day. In the tropics it was 0–5 °C warmer, with the Arctic being 20–35 °C warmer and the Antarctic being at a similar temperature to the Arctic, as opposed to being markedly colder now (Fig. 4.6). The distribution of continents in the Cretaceous was probably the major factor in maintaining these balmy conditions. The existence of a circum-equatorial seaway and the absence of circum-polar currents may explain the warmth (Fig. 4.7). Average temperatures may have been as high as 17 °C in polar regions and this would also have had a profound effect on the vertical circulation patterns in the oceans, which are now driven by the descent of cold waters in these regions (see Section 3.6), making the turnover far more sluggish. The possibility that changes in ocean circulation can have such a profound impact on the global climate is of particular interest in considering current events (see Section 8.3).

It is, however, a measure of the gaps in our knowledge that there is still dispute about just how warm this period was. There are doubts about whether the tropical oceans were as warm as is usually assumed, and arguments over the extent to which continental interiors would have experienced much colder winters than data based on ocean sediments would infer yet another example of the problems of building up a reliable picture of past climates on the basis of fragmentary information. It is also another example of the value of understanding past climates. In considering the implications of possible global warming (see Section 10.2), the extreme warmth of the late Cretaceous has been used as an analogue for examining future climatic conditions.

At the end of the Mesozoic era there was a sudden brief cooling. This coincided with the mass extinction which wiped out the dinosaurs.

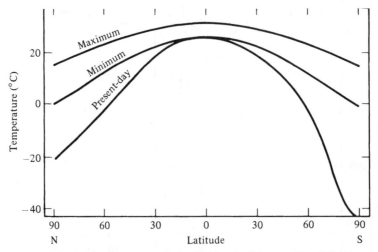

Figure 4.6 The temperature limits (maximum and minimum) with respect to latitude for the mid-Cretaceous (~100 Ma) based on the full spectrum of biological, chemical and physical observations. The limits are compared with present-day values (Brown et al., 1992, Fig. 24.8).

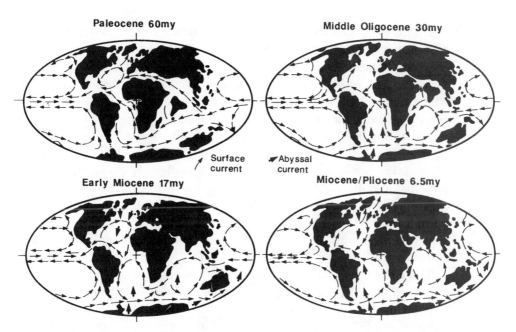

Figure 4.7 During the Cenozoic the continents moved to the positions we recognise and the climate has been strongly influenced by changes in ocean circulation. Two events matter most: the opening of the Antarctic circum-polar seaway (~25–30 Ma), and the closure of circum-equatorial seaway which was completed in the Pliocene when the Isthmus of Panama emerged (Van Andel, 1994, Fig. 11.1).

The termination of the Cretaceous is the best known example of where a cataclysmic event appears to have caused a mass extinction. The favoured explanation is that this was caused by a collision with an asteroid. The arguments for and against this theory which would have caused a huge temporary dislocation of the climate are examined in Chapter 8.

The Cenozoic era, which covers the last 65 My of the Earth's history, has experienced a long-term cooling trend (see Fig. 4.5), although initially the warmth remained on a par with the late Cretaceous. As the continents moved into the distribution we now recognise (see Fig. 4.7), things began to change. After a brief, exceptionally warm, hiccup around 55 Ma, the real cooling started and was greatest at high latitudes. It was not a smooth decline. Instead, long relatively stable periods were interspersed by more rapid cooling events at around 50 and 38 Ma. Glaciers that may have first formed on Antarctic mountains around 50 Ma appear to have grown rapidly during the cooling around 38 Ma at the end of the Eocene and beginning of the Oligocene to judge by the drop in sea level at that time. More significant were the changes in ocean circulation which occurred during the Oligocene. The Drake Passage opened up between South America and Antarctica and the movement of Australia farther north allowed the circumpolar current to develop and isolated Antarctica climatically. Around 29 Ma there was a further growth in the Antarctic ice-sheet, but there is evidence that, as late as 25 Ma, there were forested areas on the continent.

A further cooling around 12–14 Ma led to mountain glaciers starting to form in the northern hemisphere and the establishment of a permanent ice sheet over East Antarctica. The final stages of the formation of the Antarctic ice-cap occurred during a sharp cooling between 6 and 5 Ma. At the same time, ice started to form in the central Arctic Ocean. The other major climatic event of this period was the isolation and subsequent desiccation of the Mediterranean around 5.8 Ma. Known as the *Messinian Salinity Crisis*, the deposition of huge quantities of salt as the sea dried up and the water vapour entered the hydrological cycle. This would have reduced the salinity of the rest of the world's oceans and caused major perturbations of the climate. Then around 5.3 Ma tectonic activity opened the Straits of Gibraltar and a vast torrent of water flooded back into the Mediterranean sending further shudders through the climatic system.

These changes, which coincide with the boundary between the Miocene and Pliocene, included a sharp increase in sea level and a marked warming lasting until about 3 Ma in the mid-Pliocene. There was then a drop in temperature, followed by a return to relatively warm conditions. Around 2.5 Ma there was a sharp cooling and fall in sea level which is widely believed to have resulted from the initial rapid build-up of northern hemisphere ice sheets.

The sustained period of cooling during the Cenozoic era is linked to a variety of effects. Continental drift had increased the amount of land at high latitudes in the northern hemisphere making it easier for ice sheets to form. The opening and closing of ocean gateways – the Iceland-Faroe sill subsided in the late Eocene and the elevation of the Panamanian Isthmus closed the link between the Atlantic and Pacific Oceans around 3 Ma – altered ocean circulation patterns and allowed new currents to develop. The latter may have been crucial in creating the right conditions for ice sheets starting to form in the northern hemisphere.

The formation of the Himalayas and the Tibetan Plateau following the collision of the Indian subcontinent with Eurasia some 50 Ma, combined with the uplift of the Western Cordillera in North America altered the global atmospheric circulation patterns (see Section 3.2). Finally, lower levels in the concentration of carbon dioxide in the atmosphere (see Section 4.2) reinforced the cooling trend.

The Cenozoic era culminated in the Quaternary period which started around 1.6 Ma and runs up to the present time. This period of repeated glaciations, when up to 32% of the Earth's surface was covered by ice, is usually divided into the *Pleistocene*, covering the glaciations, and the *Holocene*, the warm period covering the last 10,000 years. The characteristics of these glaciations changed throughout the Pleistocene. Prior to 0.9 Ma the amplitude of variations was smaller, indicating smaller ice volume and climatic fluctuations, and higher average sea level. During the last 900,000 years the fluctuations have been larger and featured longer cold periods with greater ice volumes and lower average sea level. The reasons for this change are considered in Chapter 8.

4.2   ATMOSPHERIC COMPOSITION

Before discussing more recent climate change in greater detail, it makes sense to consider more closely some of the physical changes which occurred in parallel with the fluctuations described in Section 4.1. One obvious factor, which has been an integral part of the whole process, is changes in the level of carbon dioxide (CO_2) in the atmosphere since pre-Cambrian times. The scale of these changes is still the subject of some scientific debate. The balance between CO_2 and oxygen (O_2) is a consequence of the biotic cycle, which includes the interaction of these gases with living matter and its waste products. Oxygen in the biological cycle is produced by photosynthesis of plants and consumed in the oxidation of organic materials to form CO_2. The consumption of CO_2 by plants in the process of photosynthesis is the other half of this cycle. So changes in the productivity of the Earth's biosphere (see Section 3.5) have altered the amount of CO_2 in the atmosphere. Also, where large quantities of organic material have been deposited in the form of coal, oil and natural gas, the effect has been to deplete CO_2 levels.

On the timescale of tens of millions of years changes in the level of CO_2 in the atmosphere are, however, controlled by two other processes. First, is the atmospheric input by volcanoes. The second is the weathering of exposed silicate rocks, which depletes the atmosphere of CO_2. These two processes are linked to plate tectonics. When combined with changes in the level of plant life these processes have led to striking changes in the amount of CO_2 in the atmosphere on geological timescales. Prior to the Cambrian period, variations in CO_2 levels are the subject of great uncertainty and may have been anything from five to several hundred times current levels. These changes are, however, of great climatic importance as they were probably a central factor in the relative stability of the Earth's temperature for much of pre-Cambrian times. The lower solar constant at the time (see Section 2.2) would have led to lower temperatures but for the enhanced greenhouse effect (see Box 2.1) of the high levels of CO_2 at the time. This is why the evidence of much colder episodes (the Snowball Earth – see Section 4.1) and fluctuations of CO_2 levels are central to understanding longer term climate change.

Over the last 600 My, although CO_2 levels appear to have fluctuated less, there is considerable doubt about just how great the variations have been. In particular, there is some evidence that the huge deposition of organic material during the Carboniferous period led to major perturbations in both CO_2 and O_2 levels (Fig. 4.8). Thereafter, CO_2 levels rose, and during the Cretaceous it is estimated they were four to ten times current levels. After the mid-Cretaceous, CO_2 levels declined possibly because the sluggish

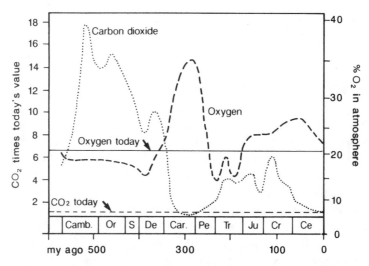

Figure 4.8 The precise levels of atmosphere constituents carbon dioxide and oxygen depends on interpreting the complex fossil record in terms of the many reservoirs, fluxes and interactions which could lock up or release carbon and oxygen from rocks. The most comprehensive analysis by Robert Berner of Yale University contains two dramatic features. First, the high levels of CO_2 in the Cambrian and Ordovician and, second, the sharp reduction in CO_2 and the build-up of oxygen in the late Carboniferous which are linked with the laying down of huge deposits of coal at the time (Van Andel, 1994, Fig. 14.6).

circulation of the oceans in a warm world created anoxic conditions at depths which allowed large quantities of carbon to be deposited in the form of organic-rich shales, which contain high concentrations of hydrocarbons. Levels of atmospheric CO_2 declined around the end of the Cretaceous, and apart from some evidence of an increase in the Eocene around 45 Ma, have remained at relatively low levels since then.

Of potentially greater interest are the changes which have occurred during the Pleistocene. Analysis of air trapped in bubbles in the ice in Antarctica (see Section 6.4.2) show changes in the concentration of trace gases (e.g. CO_2 and CH_4) which closely track changes in temperature. Although these changes are small compared with changes in CO_2 over geological time, they have important implications for explaining the physical processes driving both the last ice age and for predicting the consequences of the present-day build-up of greenhouse gases in the atmosphere (see Section 10.2).

Although vitally important in terms of evolution, changes in oxygen levels during the Phanerozoic have been of little direct consequence to climate change. Oxygen first appeared in the atmosphere in very low levels around 2,000 Ma and rose to some 10% of current levels at about 700 Ma. During the Cambrian it reached concentrations comparable with current levels. The fluctuations over the last 600 My shown in Fig. 4.8 remain the subject of scientific dispute and investigation.

4.3   SEA LEVEL FLUCTUATIONS

Rises and falls in sea level are an integral part of climate change. During the ice ages a large amount of water was locked up in the ice sheets that covered the northern continents. So the level of the oceans dropped by more than 100 m. But for much of the history of the Earth there have been no ice caps, so the question arises as to whether evidence of sea-level variations contain significant climatic information. Over the last 600 My there have been substantial long-term variations. While absolute figures are hard to produce, the scale of changes fall in the range from being as much as 200 m lower than the current level to being over 250 m above this. These fluctuations provide additional information about past climates but also highlight how climatic factors are tangled up with many other physical processes. As a consequence, care is needed to avoid jumping to conclusions about what part the climate played in these changes.

The most comprehensive analysis of changes of sea level have been prepared by geologists for the Exxon Corporation, and the resulting record is often known as the *Exxon curve,* or the *Vail curve* after Peter Vail, the leading geologist in the work. This analysis is the subject of fierce debate among geologists because of some of the conclusions reached, and because much of the data is commercially confidential and so cannot be the subject of independent scientific assessment. The reason for secrecy is that the analysis looks at the way sediments were laid down along the shores of ancient seas and how different layers show evidence of rising or falling sea levels (Fig. 4.9). Because these sedimentary layers may contain hydrocarbon reserves they are of great commercial value and oil companies are reluctant to share the data with other scientists.

In spite of the unresolved argument over the Exxon curve the broad features are trustworthy (Fig. 4.10). The first order changes involve timescales that are so long that they are unlikely to be the product of climatic factors. Although the lower sea level at the end of the Carboniferous and during the Permian may be partially due to the glacial epoch at this time, the most likely explanation for the long-term rise and fall is that it is the consequence of continental drift and changes in the rate of formation of mid-ocean ridges. Superimposed on this broad pattern is evidence of more rapid supercycles and cycles (see Fig. 4.10). More striking is the nature of the rises and falls, with the rises being relatively gradual and the falls being precipitate, giving the Exxon curve a characteristic 'saw-tooth' appearance. The intriguing question is whether these fluctuations and, in particular the sudden drops in sea level, are evidence of the formation of ice caps.

There are a considerable number of sudden substantial drops in sea level. While the timing and scale of the early examples are hard to pin down, there is considerable evidence of their existence. Little is known about the causes of the drops in sea level prior to around 300 Ma, although

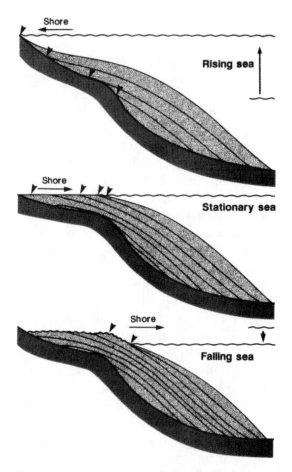

Figure 4.9 Many estimates of past sea levels depend on the interpretation of seismic studies of identifiable layers of sediments laid down on the floors of ancient oceans. The highest point at which any layer is found (see arrow) will mark the highest sea level when it was formed. Depending on whether the sea was rising, stationary or falling, the extent of a sequence of layers formed near continental margins will provide information on changes in sea level. This method is known as *sequence analysis* (Van Andel, 1994, Fig. 9.7).

those at the end of the Silurian and during the Carboniferous may be linked to ice sheet formation. In the late Permian there was a sharp drop around 255 Ma, which may be linked to the mass extinction at this time (see Section 5.3). Then in the Triassic there was an event at about 232 Ma, but after that there are no striking examples during the next 100 My. The next two major falls are at about 128 and 126 Ma and coincide with the longer term low sea level in the early Cretaceous. These may be linked with continental glaciation (see Section 4.1). Thereafter, in spite of the broad rise in sea level, there were major short-term dips in the late Cretaceous at around 90 and 67 Ma. During the Cenozoic the detailed picture becomes more clear. There were short sharp falls at about 58.5 and 49.5 Ma, followed by

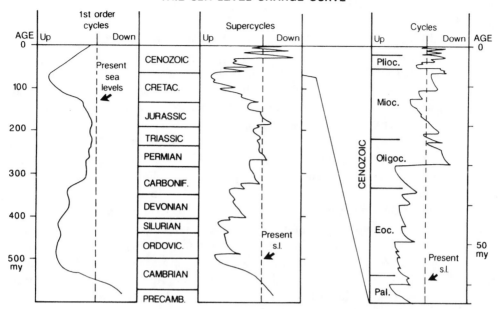

VAIL SEA LEVEL CHANGE CURVE

Figure 4.10 Sequence analysis of global sea level changes during the Phanerozoic on three different scales. The first order changes show the underlying megacycles of several hundred My in duration. Superimposed on these long-term changes are a series of 'supercycles' whose length varies from 20 to 50 My in the Palaeozoic to 4 to 15 My in the Mesozoic and Cenozoic, while more detailed analysis of the Cenozoic shows an increasing number of shorter 'cycles' (Van Andel, 1994, Fig. 9.8).

a series of falls and rises of levels between 39.5 and 35.5 Ma, and again between 30 and 25 Ma. The fall at 30 Ma appears to have been particularly marked and may be linked to the start of the build-up of the Antarctic ice sheet (see Section 4.1), but the scale of the drop suggests other wider tectonic effects were also involved. There were then two sharp dips around 16.5 and 15.5 Ma. Since around 10.5 Ma there have been a series of low levels (~ -100 metres), punctuated with higher stands, with the lowest levels at 2.5, 1.5 and since 0.9 Ma. Changes during the last ice age are considered in Section 4.4, while those over the last 20,000 years are discussed in Section 5.4.

Overall, data on sea level changes charted in the Exxon curve imply a more dramatic and variable global environment than is normally inferred from other geological records. As noted earlier, however, it is not possible to make a direct connection between rises and falls in sea level and climate change, because other factors could have been the driving force for both sets of variations. So factors such as tectonic activity, including both volcanism and continental drift, and fluctuations in ocean chemistry leading to changes in atmospheric composition have to be considered in seeking to explain the observed changes (see Chapter 8).

4.4   THE ICE AGES

Although there have been a number of epochs when large areas of the Earth were covered by ice sheets, the term Ice Age is most frequently applied to the Pleistocene glaciations. Moreover, how the discovery of these icy events has come to be associated with the Swiss naturalist Louis Agassiz, is an excellent example of how geological evidence was put together to provide a coherent theory of climate change.

The possibility of ice sheets covering part of northern Europe had first been proposed by James Hutton, the founder of scientific geology in 1795, and reiterated by a Swiss civil engineer, Ignaz Venetz in 1821. In 1824, Jens Esmark, a Norwegian geologist, offered the theory that Norway's mountains had been covered by ice, while in 1832 a German professor of forestry, Bernhardi, published a paper suggesting that a colossal ice sheet extended from the North Pole to the Alps. But these ideas received scant attention.

In the summer of 1836, while on a field trip in the Jura Mountains with Jean de Charpentier, a friend of Ignaz Venetz, Agassiz became convinced that blocks of granite had been transported at least 100 km from the Alps (Fig. 4.11). In 1837 he first coined the term Ice Age (*die Eiszeit*) and in 1840 his proposals were published in a ground-breaking book. At first the theory was ridiculed by the geological community, but his passionate advocacy of the ice age was to prevail.

Agassiz travelled to Scotland where he saw more evidence of glaciation and then in 1846 arrived in Nova Scotia where again the evidence of ice was plain to see. In 1848 he joined the Harvard faculty and was active in many fields, notably marine science, but continued glacial research in New England and around the Great Lakes. Over the next few decades a variety of geological evidence made it clear that many features of the northern hemisphere could only be explained by ice ages and Agassiz was vindicated, although in some quarters the subject remained controversial until the end of the nineteenth century.

Following the work of Agassiz an orderly view built up during the remainder of the nineteenth century about ice ages. This was that over the last six hundred thousand to a million years there had been four glacial periods lasting around 50,000 years separated by warm interglacials ranging from 50,000 to 275,000 years in length. The present interglacial started about 25,000 years ago and was destined to last at least as long as previous interglacials and possibly indefinitely. The principal evidence of the stately progression was seen in glaciated landscapes of the northern hemisphere. The scoured U-shaped valleys, eroded mountains, *drumlins* (mounds of stiff boulder clay moulded under, and by, the creeping ice), *eskers* (ridges of gravel and sand formed by the meltwater flowing out of edges of the ice sheets), large glacial boulders (erratics), glacial tills and terminal moraines are obvious landscape features. This view was encapsulated in the work of Penck and Bruckner published between 1901 and 1909.

Figure 4.11 Glacially polished rocks and morainic debris at the edge of the Zermatt glacier. From Louis Agassiz's book *Études sur les glaciers* published in 1840 (*Cambridge Encyclopaedia of Earth Sciences*, Fig. 1.12).

While some other geologists were more cautious about the chronology of the events, the broad picture of four major glaciations was the accepted interpretation of the geological evidence in both Europe and North America. The nomenclature for these ice ages and the associated interglacials, which reflects the local sites where the evidence of their existence was identified, together with the broad timing of these events is summarised in Table 4.1.

In the 1950s, the situation started to change. Caesari Emiliani, at the University of Chicago, published a set of papers on the properties of fossil shells of the tiny creatures found in the sediments of the tropical Atlantic and Caribbean. Using the reversal of the Earth's magnetic field 700,000 years ago as a marker, he was able to show that there had been seven glacial periods since then, occurring every 100,000 years or so (Fig. 4.12). This new picture of the most recent ice ages was not immediately accepted. But, since the 1960s a growing body of evidence principally from ocean sediments (see Section 6.4.3), but also from pollen records from part of Europe which had not been covered by ice, and Antarctic ice cores has confirmed Emiliani's conclusions. It is now agreed that glacial periods have occurred more frequently during the Pleistocene than early theories suggested and, during the last 800,000 years their period has been roughly every 100,000 years. These cold periods have been interspersed by shorter warm interglacials (see Fig. 4.12). Within each glacial period there were substantial fluctuations in the climate ranging from extreme cold to

TABLE 4.1 CORRELATION OF PLEISTOCENE STAGES

Stages	Date (kyBP)[a]	Alps	Northern Europe	European Russia	North America
Post-Glacial	0–25	Post-Glacial	Flandrian	Post-Glacial	Post-Glacial
4th Glacial	25–75	Würm	Weichsel	Valdai	Wisconsin
Interglacial	75–125	Riss–Würm	Eem	Mikulino	Sangamon
3rd Glacial	125–175	Riss	Saale	Moskva	Illinois
Interglacial	175–450	Mindel–Riss	Holstein	Likhvin	Yarmouth
2nd Glacial	450–500	Mindel	Elster		Kansan
Interglacial	500–600	Günz–Mindel	Cromer	Morozov	Afton
1st Glacial	600–700	Günz	Menap	Odessa	Nebraskan

[a] These dates and the correlations between the events identified in different parts of the world must be regarded as tentative in the light of more recent geological evidence (see later). Nevertheless, these names are still often used to describe past climatic events and so it is as well to know to what they refer.

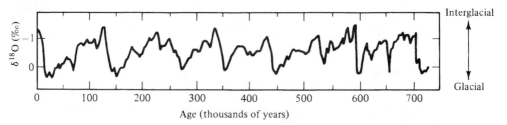

Figure 4.12 An example of the analysis of oxygen isotopic composition of planktonic foraminifera measured in deep sea cores. This curve shows the waxing and waning of the ice ages and the characteristic 'saw-tooth' behaviour as the ice sheets build up relatively slowly and then collapse more rapidly (Brown et al., 1992, Fig. 24.2).

near-interglacial warmth. Statistical analysis of these fluctuations shows that they are dominated by three major cycles of periods of around twenty-one-, forty-one- and one hundred thousand years (see Section 2.1.4).

More detailed analysis of both ocean sediment and ice-core data for the period covering the last glacial shows the striking suddenness of some of the changes that occurred (Fig. 4.13). In particular, the dramatic warmings which can be seen in the Greenland ice core appear to coincide with what are known as *Heinrich layers* in the ocean sediment data (see Table 4.2). These layers, which are named after the scientist who first identified them, are thought to have been produced by debris carried out into the North Atlantic by a surge of icebergs resulting from the sudden collapse of part of the ice sheet covering North America. These sudden changes raise important questions about how such influxes of freshwater might have altered the deepwater circulation of the North Atlantic (see Section 8.3). So the

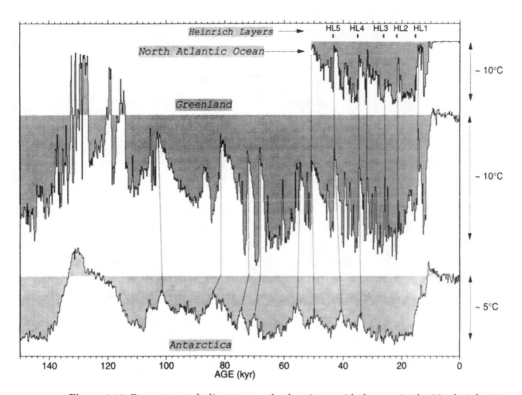

Figure 4.13 Reconstructed climate records showing rapid changes in the North Atlantic and in Greenland; the corresponding events (indicated by thin dashed vertical lines) are damped in the Antarctic record. Temperature changes are estimated from the isotopic content of ice (Greenland and Antarctica) and from faunal counts (North Atlantic). HL1 to HL5 indicate sedimentary 'Heinrich' layers (IPCC, 1995, Fig. 3.22).

last glacial was not a period of unremitting cold, but exhibited marked fluctuations on timescales from years to millennia.

The impact of ocean sediment data shows how new evidence can transform our view of the past. A century of work on glacial deposits has shown that reliable chronologies cannot be achieved for the continents because deposition was not continuous, and because of lack of fossils led to undateable sequences. In contrast, the marine record is often continuous and more easily dateable, and while the rate of deposition of sediments may fluctuate, the possibility of getting a complete measure of changes of climatically sensitive parameters becomes possible (see Section 6.4.3). So, only after it became technically feasible to retrieve cores of ocean sediments and measure their properties accurately, was it a practical proposition to put all of the continental evidence into context by correlating ocean sediments with those on land. The subsequent development of ice core measurements (see Section 6.4.2) was then able to reinforce the conclusions emerging from deep-sea work.

The product of all this work is an increasingly ordered picture of the waxing and waning of the ice sheets across the northern hemisphere during

TABLE 4.2 THE CHRONOLOGY OF EVENTS DURING THE LAST ICE AGE

Δ^{18}O Stage[a]	SPECMAP age (kyBP)	Heinrich events number (kyBP)	Comments (onset kyBP)
1	0–15	H1 (16.5)	Ice sheets start to retreat (17 kyBP)
2	14–24	H2 (23)	Maximum extent of ice sheets (25–18 kyBP)
3	28–59	H3 (29)	Interstadial in N America and Europe (30 kyBP)
		H4 (37)	Fennoscandian ice sheet expanded (42–35 kyBP)
		H5 (51)	Partial amelioration in Europe (45–40 kyBP)
			Much of W Canada not glaciated (60–30 kyBP)
4	59–74	H6 (~70)	Interstadial in Europe (60 kyBP)
			Considerable doubt about extent of glaciation in North America (79–65 kyBP); cool dry conditions in northwest US
			Expansion of Fennoscandian ice sheet (80–75 kyBP)
5a	74–85		Pronounced interstadial (85–80 kyBP); return of cool mixed forest to parts of northern Europe. Warmer wetter summers northwest US than now
5b	85–94		Northern Europe descends into taiga and tundra. Cool dry conditions in northwest US
5c	94–107		Pronounced interstadial (105–94 kyBP), forests return to northern Europe
			Warmer wetter summers northwest US than now
5d	107–117		Onset of glacial climate with northern Europe shifting from temperate forest to taiga
			Cooler and wetter in northwest US than now
5e	117–130		The last interglacial (the Eemian), generally warmer and drier in North America and Europe than now

[a] The numbers denote alternately cold (even stages plus 5a, 5c and 5e) and warm periods (odd stages plus 5b and 5d).

the last ice age. These fluctuations are summarised in Table 4.2. A key element in this presentation is the agreed stages in the chronology from the oxygen isotope record in the ocean sediments, which is known as SPECMAP (see Section 6.4.3). The conditions on the northern continents after the major build-up between 117 and 74 kyBP is based on fragmentary evidence. This reflects that as the ice sheets built up, and then fluctuated with changing climatic signals, their extent would have varied appreciably from place to place and time to time around the northern hemisphere. So there is considerable uncertainty about whether these changes were synchronous or were independent of one another. Table 4.2 is therefore designed to give some idea of where there is broad agreement about general developments.

The last glacial period reached its greatest extent around 18,000 years ago. At this time, ice sheets up to 3 km thick covered most of North America, as far south as the Great Lakes, all of Scandinavia and extending to the northern half of the British Isles and the Urals. In the southern hemisphere much of Argentina, Chile and New Zealand were under ice, as were the Snowy Mountains of Australia and the Drakensbergs in South Africa. The total amount of ice locked up in these ice sheets has been estimated to be between eighty-four and ninety-eight million cubic kilometres as compared to the current figure of about thirty million cubic kilometres. This was sufficient to reduce the average global sea level by between 90 and 120 m.

The global average temperature at the height of the last ice age was at least 5 °C lower than current values. Over the ice sheets of the northern hemisphere the cooling was around 12–14 °C. The position in the tropics is less clear. A major international study (CLIMAP) in the 1970s concluded that temperatures were much closer to current figures and over some of the tropical oceans there might even have been warmer conditions. More recent studies suggest parts of the tropics might have been as much as 5–6 °C cooler, although an average figure of some 3 °C cooler with the overall global temperature being some 6–8 °C lower. These uncertainties pose a considerable challenge to the computer models used to predict global warming (see Chapter 9), as they tend to show greater changes in temperature at higher latitudes than in the tropics and have difficulties in reproducing these ice age conditions. What is certain is that around 15,000 years ago a dramatic warming started and, with it, our perspective alters.

4.5   THE END OF THE LAST ICE AGE

So far the evidence of climate change has considered timescales that are beyond human experience. The consequences of the huge, but gradual changes over geological time seem too remote to influence our lives.

By this criterion, we might assume that the emergence of the Earth's climate from the last ice age over the period of 15,000–10,000 years ago was also of little immediate relevance to current concerns about climate change. Not so: the eventful transition between glacial and post-glacial climates prior to the settled period covering the last 10,000 years (the *Holocene*), is of particular importance to understanding the stability of the climate.

The erratic retreat of the ice sheets at the end of the last ice age and the associated advance of vegetation and forests left many clues on the landscape. The terminal moraines of the ice sheets together with the fluctuating levels of the lakes trapped behind the ice provided plenty of evidence for geologists to identify various stages of the retreat in both northern Europe and North America. At the same time botanists extracted information from both the vegetation and insects collected in peat deposits, plus pollen records from both peat beds and the sediments of lake beds. During the early part of this century this work produced detailed but somewhat separate chronologies for events in Europe and North America. The essence of this work is reviewed in Table 4.3. These presentations make use of more recent measurement techniques, notably Greenland ice-core data (see Section 6.4.2) to place events on a consistent timescale.

The most important feature of the changes detected is the sudden switches that occurred, most notably in respect of the onset of the Bølling interstadial and the beginning and end of the Younger Dryas event. First identified in pollen records and confirmed by measurements from the Greenland ice cores, these dramatic shifts in various meteorological parameters show that rapid changes in circulation patterns occurred within a few years. These were a feature of both the end of the last ice age and during it (see Fig. 4.13). Even more controversial is the possibility that such sudden changes also occurred during the last interglacial (the Eemian between 117,000 and 130,000 years ago), but there is considerable doubt about the reliability of the ice core (see Section 6.4.2). The relevance of these findings to the current debate on climate change are explored in Section 8.3. The essential point is that the evidence from the end of the last Ice Age shows the climate is capable of sudden large shifts: ice ages are not a matter of glacial slowness. What is more, unless it can be shown that this behaviour can only occur during and around the end of glacial periods, it may be of real relevance to current events.

Although, when the changes came, climatic patterns shifted into a much warmer regime swiftly, there were a number of odd features. Around Antarctica there is little evidence of the dramatic ups and downs that marked the emergence of the northern hemisphere from the icy grip of the glacial. More striking is that the warming started two to three thousand years earlier in the southern hemisphere. This delay may be related to the presence of the huge ice sheets on the northern continents, which exerted an inertial influence at higher latitudes for several millennia. This is hardly

TABLE 4.3 CHANGES AT END OF THE LAST ICE AGE

Time (BP)	Northern Europe	North America
17,000	Glaciers retreat in Alps	Major collapse of part of ice sheet
16,000		Wet period across southern US with lake levels from Great Basin to Florida reaching the highest levels around 15,000 BP
15,000	Abrupt warming starts ~14,900 BP with evidence of forest expansion northwards (known as Bølling interstadial)	
14,000	Return of cold conditions around 14,000 BP for as much as 500 years (known as Older Dryas)	Cordilleran ice sheet retreats from Puget Sound
	Warming around 13,500 BP (known as Allerød interstadial) although some doubt as to whether as warm as Bølling	
13,000	Around 12,900 BP a sudden and dramatic cooling (known as Younger Dryas), forests retreated far south, and ice sheets expanded	Drainage of meltwater from Laurentide ice sheet switches from Gulf of Mexico to St Lawrence
		Together with surges of icebergs out of Hudson Bay these changes could have cooled the North Atlantic, thereby triggering the Younger Dryas
12,000	Around 11,600 BP sharp and sustained warming, the initial stages of which are usually known as the Pre-Boreal	Readvance of the Laurentide ice sheet may have redirected meltwater back to Gulf of Mexico and so switched off the Younger Dryas
11,000	Forests started to spread across northern Europe	Cordilleran ice sheet melts rapidly and disappears at around 10,000 BP
10,000	Forests became established at high latitudes, although the Fennoscandian ice sheet did not finally disappear until around 8500 BP	Laurentide ice sheet declines more slowly and even readvances (see Table 4.4), and finally melts away at around 7000 BP

surprising if you think about how long it took for the ice to melt. Even allowing for the the dramatic changes in atmospheric circulation that may have occurred in the northern hemisphere, together with the increased summer insolation at high latitudes, it would have taken several thousand years for the ice to melt.

So, until around 10,000 years ago the global climate was adjusting to a series of huge climatic changes. Thereafter, the climate settled into what looks like an extraordinarily quiet phase, when viewed against the upheavals that went before. Apart from a short-lived sharp cooling of about 1.5–3 °C around the North Atlantic region between about 8,400 and 8,000 years ago, the Holocene has been climatically benign compared with the preceding 100,000 years. This temporary setback was probably linked to the sudden release of the floodwaters of two lakes which had collected over central Canada behind the melting Laurentide ice sheet. This surge of freshwater out through the Hudson Strait and into the Labrador Sea could have disrupted the circulation of the North Atlantic and precipitated one last cold event.

Over northern Europe, the Fennoscandian ice sheet largely disappeared by around 8,500 years ago after creating a series of meltwater lakes in the region of the Baltic. The expansion of forests reached their northern limits across Eurasia around 7,000 years ago, some two- to three-hundred kilometres farther north than their current extent. In northern Canada this expansion was delayed a further two- to three-thousand years, by the slower collapse of the Laurentide ice sheet, which did not completely disappear until around 6,000 years ago.

4.6   THE HOLOCENE CLIMATIC OPTIMUM

By around 6,000 years ago post-glacial warming reached a peak. On the basis of evidence of tree cover the average summer temperature in mid latitudes of the northern hemisphere was 2–3 °C warmer (Fig. 4.14). Not only had trees spread farther north than now but extended to higher into upland areas: in the British Isles trees grew at levels 200 to 300 m above the current timberline. This phase of the climate is generally known as the 'Atlantic' period, because it is assumed to have featured a strong westerly circulation which brought warm wet conditions to high latitudes of the northern hemisphere. As such it is sometimes seen as an example of what we might expect if the current global warming (see Section 4.10) were to continue.

At lower latitudes the warmer conditions showed up most noticeably in a stronger summer monsoon circulation. This brought heavier rainfall to many parts of the sub-tropics. This affected not only the Indian sub-continent but also the Middle East and across much of the Sahara. Ice cores from glaciers high in the Peruvian Andes confirm that the period

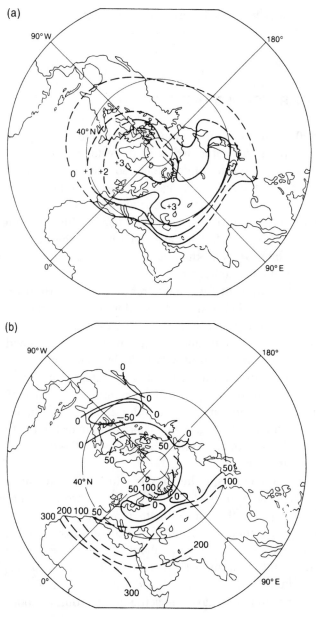

Figure 4.14 Departures of: **(a)** summer temperature (°C), and **(b)** annual precipitation (mm), from modern values for the Holocene climatic optimum around 6000 BP (IPCC, 1990, Fig. 7.5).

from 8,000 to 5,000 years ago marked the climatic optimum in the tropics. This warmer, wetter period probably provided the right conditions for the spread of agriculture from its earliest beginnings in the 'fertile crescent' of the Near East some 10,000 years ago. But, the evidence of the unusual stability of the climate during the Holocene period (see Fig. 4.13) has the most profound implications for assessing the potential consequences of current climate change.

4.7   CHANGES IN RECORDED HISTORY

Around 5,500 years ago the climate began to cool gradually and become drier. This trend coincided with the establishment of the ancient civilisations of Egypt and the Middle East. So from this point onwards the physical evidence of climate change can be combined with historical records from various civilisations. But, as already noted, the changes since around 10,000 years ago have been small compared with the sudden shifts at the end of, and during the last ice age. This means the evidence is much more difficult to interpret, and the fluctuations in human activities have, if anything, served to confuse matters.

The clearest example of the broad trend since the Holocene climatic optimum is the desiccation of the Sahara. More generally there is evidence of a decline in rainfall in the Middle East and North Africa setting in around 4,000 years ago. At the same time the treeline across the Canadian Arctic and Siberia started to recede southwards. This trend continued until near the end of the first millennium AD. What is less clear is how great the fluctuations on this broad trend were. The considerable evidence of mountain glaciers around the world expanding around 2500 BC, but then receding to high levels around 2000 BC (Fig. 4.15). Thereafter, over the next few centuries, they expanded again, apparently in phase with the desiccation around the Mediterranean.

The most dramatic effects of the cooling which led to the advance of the glaciers after 2000 BC appears to have occurred across northern Europe. Tree-ring data suggests that the climate remained warm and dry until around 1700 BC, and then cooled until the fifteenth-century BC, at which point there was a marked warming for about 70 years, before an even colder wetter period set in from around 1400–1230 BC. This led to some glaciers in the Alps expanding to limits not matched since, and there is evidence of widespread abandonment of lakeside settlements around 1200 BC as water levels rose.

There is plenty of evidence of glaciers around the world expanding again around 200 BC and between AD 500–800. Tree-ring data from northern Eurasia and the White Mountains in California (Fig. 4.16) show cooler conditions prevailed during the latter period. Changes in temperature and rainfall at lower latitudes during these times are less easy to identify. Furthermore, attempts to link these, and earlier changes to the 'Dark Ages' when ancient civilisations went into decline is the subject of heated debate (see Section 5.4). For ease of analysis, a summary of the various changes during the period 10 000 BC and AD 1000 is given in Table 4.4.

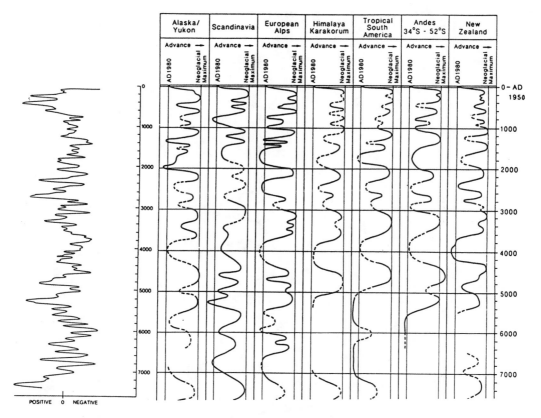

Figure 4.15 An analysis of the fluctuations of glaciers in the northern and southern hemispheres during the last 7,600 years, compared with radiocarbon production variations. Periods of negative radiocarbon production signal lower solar activity and are associated with glacial advance, which is a sign of a cooling climate (Burroughs, 1994, Fig. 4.10).

What is more clear is that there were shorter periods of climatic deterioration, which may or may not have coincided with the decline of certain civilisations. The identification of these events, their causes and the question of their historical impact are considered in later chapters. As a general observation, historical records provide relatively little insight. While there are plenty of examples of custom and practice (e.g. the extent of the production of olive oil and wine, the construction of bridges and the design of buildings), which can be used to infer that climate change had a part to play in events, their interpretation is the subject of dispute. Specific references to the weather are available here and there. So for instance a compilation of the examples referring to Italy between around 300 BC and AD 1300 suggest warmer, drier conditions in the third to fifth centuries AD, and again in the eighth and tenth centuries AD, with wetter conditions at other times. But, apart from the occasional tantalising clues like the weather diary of Ptolemy which indicates that between AD 127 and 151 the climate of Alexandria was much wetter than now, the historical

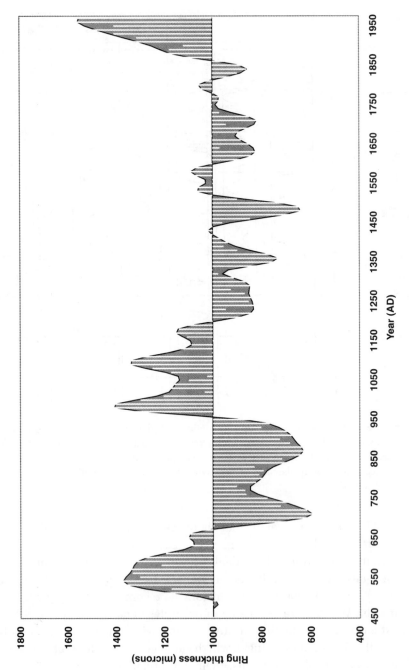

Figure 4.16 Ring widths (twenty-year averages) from bristlecone pines growing high in the White Mountains, California, from AD 470. The variations in tree-ring width at this altitude are an indication of summer warmth and/or duration of growing season (data from Tree-ring Laboratory, University of Arizona).

TABLE 4.4 CHANGES DURING THE HOLOCENE (8000 BC– AD 1000)

Period[a]	Central England	Europe	Sahara	North America
Sub-Atlantic 900 BC onwards	15.1/4.7/9.3[b] (900–450 BC)	Cooler, wetter 900–450 BC, then warmer and drier until ~AD 400 then colder and more variable until ~AD 1,000	Generally arid but some evidence of more winter rain in North Africa	Northern Great Plains drier than now until AD 1200; with extreme droughts AD 200–370, 700–850 and 1000–1200. White Mountains coldest ~AD 900
Sub-boreal 1000–3000 BC	16.8/3.7/9.7	Generally dry and warm until ~1500 BC, much cooler and wetter until ~1200 BC, then warmer	Moist period ended ~2350 BC thereafter much drier	Cooling in White Mountains ~1200 BC Northern Great Plains warmer (~1–2 °C) and drier until ~2000 BC
Atlantic 3000–5500 BC	17.8/5.2/10.7 wetter than present	Rapid peat build-up in N. Europe, but cold spell in Alps 5500–4500 BC	Lake Chad still 30–40 m higher than present, moist except 3500–3800 BC	Sharp cooling in White Mountains 3300–2800 BC
Boreal 5500–6900 BC	16.3/3.2/9.3	~1–2 °C warmer	Wetter in Egypt	Laurentide ice sheet expanded (Chochrane readvance) 4700–6500 BC
Pre-Boreal 6900–8300 BC			Lake Chad 52 m higher (area 400,000 km^2)	

[a] The dates quoted for these periods, whose titles are widely and flexibly used to attach a coherent timescale to evidence of climatic change, vary considerably. Here a consensus view is taken on the timing of these events.

[b] The three figures are quoted for Central England temperature – high summer (July and August), winter (December to February) and the annual value – based on biological evidence. The values for the twentieth century are 15.8/4.2/9.4.

records are unable to provide any clear evidence as to just how much the climate varied between 3000 BC and AD 1000.

4.8   THE MEDIEVAL CLIMATIC OPTIMUM

Although there is no sudden change in the amount of historical evidence on the weather at around AD 1000, there are more and more pointers to the fact that the climate of northern Europe and around the North Atlantic became warmer during the ninth and tenth centuries. In part, this is inferred from the expansion of economic and agricultural activity throughout the region. Grain was grown farther north in Norway than is now possible. Similarly, crops were grown at levels in upland Britain, which has proved uneconomic in recent centuries. The Norse colonisation of Iceland in the ninth century and of Greenland at the end of the tenth century are also seen as clear evidence of a more benign climate.

There is little doubt the warmer conditions over much of northern Europe extended into the eleventh and twelfth centuries, but the evidence from other sources provides a complicated picture. Tree-ring data prepared by Keith Briffa of the Climatic Research Unit, University of East Anglia and other researchers (see Section 6.4.1) from northern Fennoscandia shows there was a warm period between 870 and 1100. These data also show a warm period around 1360–1570. However, similar measurements made on trees growing in the northern Urals show no evidence of the earlier warm period. Instead, the warmest periods are in the thirteenth and fourteenth centuries together with the late fifteenth century.

These regional variations are found in other records. The Greenland ice-core data (see Section 6.4.2) shows that warmer conditions started earlier, around AD 600, and reached a peak at the beginning of the twelfth century before showing a marked decline in the next 200 years (Fig. 4.17). Other records do not, however, show any evidence of warm conditions, which could be linked to this warm period. So the overall picture lacks coherence, which may reflect the limited geographical coverage of the records and the fact that they relate only to some seasons. This means that it is not possible to conclude, as some commentators have, that global temperatures in the Medieval Climatic Optimum were comparable to the warm decades of the late twentieth century.

4.9   THE 'LITTLE ICE AGE'

By comparison with the uncertainties of preceding centuries, the evidence of the cooler period between the mid-sixteenth and mid-nineteenth centuries appears to be built on firmer foundations. It is the best known example of climate variability in recorded history. The popular image is of frequent

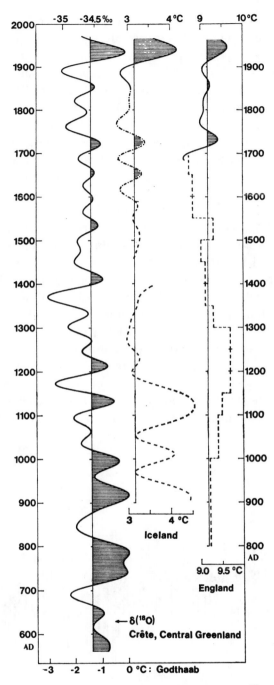

Figure 4.17 Comparison between changes in the ^{18}O concentration in a Greenland ice core, which has been calibrated in terms of the annual temperature at Godthaab, and temperature records from England and Iceland. The data has been smoothed using a sixty-year low-pass filter (see Section 7.3) except for the dashed record for England. The dashed curves are based on indirect evidence and the dashed dot section of the Icelandic record on systematic ice observations (Robin, 1983, Fig. 6.9).

cold winters with the Frost Fairs on the Thames in London. Elsewhere in Europe the same image of bitter winters prevails together with periods when cold, wet summers destroyed harvests. Widely known as the *Little Ice Age*, the period has been closely studied by climatologists for many years. This growing body of work shows that, as with all aspects of climatic change, the real situation is more complicated than the simple stereotype suggests.

There is no doubt that the climate in Europe deteriorated in the second half of the sixteenth century. The glaciers in the Alps expanded dramatically and reached advanced stages at the end of the 1590s. Studies of a wide range of historical records of both the weather and related behaviour of crops, and flora and fauna by Christian Pfister at the University of Bern clearly show that in Switzerland the period 1570–1600 featured an exceptional number of cool wet summers (Fig. 4.18). They also confirm that the winters of the 1590s were particularly severe, while the Ladurie wine harvest dates show the poor summers of the late sixteenth century (Fig. 4.19).

The interpretation of the Little Ice Age as a period of sustained cold breaks down in the subsequent decades. The growing seasons returned to more normal conditions after the 1590s, although the incidence of cold winters remained high. Records kept by Dutch merchants of when the canals were frozen and trade interrupted show that from 1634 until the end of the seventeenth century the winters were roughly 0.5 °C colder than in subsequent centuries, with the winters of the 1690s being particularly severe. The 1690s stand out more generally as being exceptionally cold with frequent harvest failures, notably in Finland, France and Scotland (see Section 5.5).

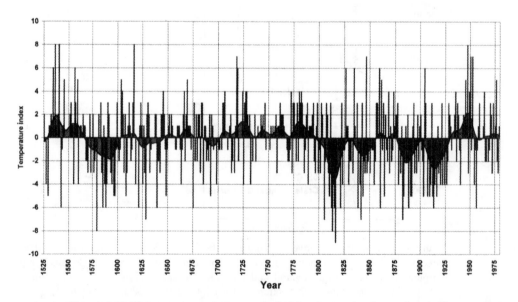

Figure 4.18 The summer temperature index for Switzerland, together with smoothed data showing longer term fluctuations (data taken from Pfister's paper in Bradley & Jones, 1995) (Burroughs, 1997, Fig. 2.9).

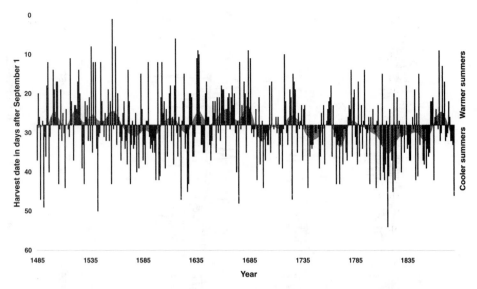

Figure 4.19 The date of wine harvests in northern France and adjacent regions, together with smoothed data showing longer term fluctuations (data taken from Le Roy Ladurie and Baulant, 1980) (Burroughs, 1997, Fig. 2.6).

Instrumental records for Central England Temperature (CET) started in 1659 and the records for de Bilt in the Netherlands in 1705 (see Section 6.1) provide more detail. The CET series confirms the exceptionally low temperatures of the 1690s and, in particular, the cold late springs of this decade. Equally striking is the sudden warming from the 1690s to the 1730s (Fig. 4.20). In less than forty years the conditions went from the depths of the Little Ice Age to something comparable to the warmest decades of the twentieth century. This balmy period came to a sudden halt with the extreme cold of 1740 and a return to colder conditions, especially in the winter half of the year. Thereafter, the next 150 years or so do not show a pronounced trend. Various other series for other European cities from the mid-eighteenth century onwards confirm this conclusion. Although the winters begin to become noticeably warmer from around 1850, the annual figures did not show any appreciable rise until well into the twentieth century. Furthermore, the temperatures for the growing season (April to September) remained virtually steady until the present day (see Fig. 4.20). The emergence from the Little Ice Age, such as it was, in Europe was concentrated in the winter half of the year.

A more striking feature is the general evidence of interdecadal variability. So, the poor summers of the 1810s are in contrast to the hot ones of the late 1770s and early 1780s, and around 1800. The same interdecadal variability shows up in more recent data. The 1880s and 1890s were marked by more frequent cold winters, while both the 1880s and 1910s had more than their fair share of cool wet summers.

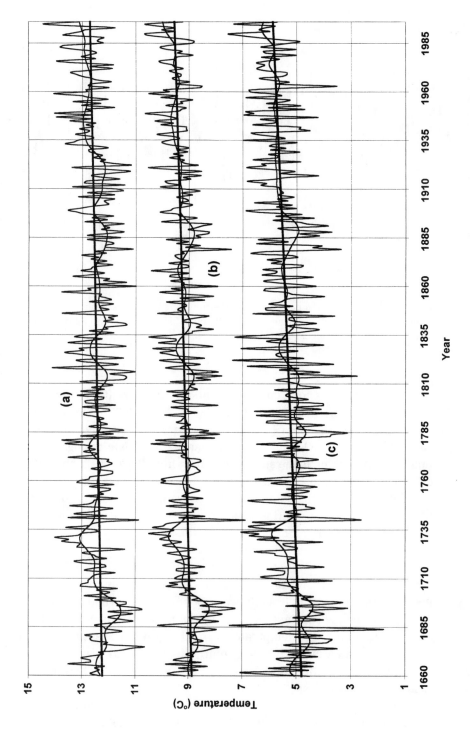

Figure 4.20 Temperature records for Central England since 1659 showing how most of the increase in the annual temperature **(b)** has been the result of warming in the winter half (October to March) of year **(c)**, as opposed to relatively little change in the summer half (April to September) of the year **(a)**.

A similar variable story emerges from other parts of the world. In eastern Asia the seventeenth century was the coldest period, with another set of cold decades around 1800. But the rest of the nineteenth century does not show the frequent periods with low temperatures seen in Europe. Conversely, the North American records show that the coldest conditions were in the nineteenth century. Tree-ring data suggest that the seventeenth century was also cold in northern regions, but in the western United States this period appears to have been warmer than the twentieth century. Limited records for the southern hemisphere indicate that the most marked cool episodes occurred earlier, principally in the sixteenth and seventeenth centuries. Evidence from the tropics is even sparser, but measurements of the ice cores from the Peruvian Andes (Fig. 4.21) suggest that there was a

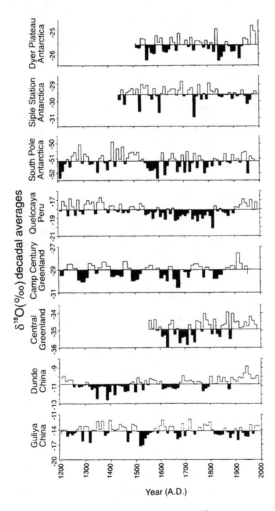

Figure 4.21 Decadal averages of the $\delta^{18}O$ records from ice cores in several widespread locations. The shaded areas represent isotopically more negative (cooler) periods relative to the individual record means (IPCC, 1995, Fig. 3.21).

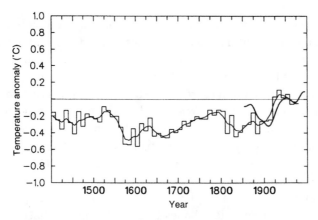

Figure 4.22 Decadal summer temperature index for the northern hemisphere up to 1970–79. The record is based on the average of sixteen proxy summer temperature records from North America, Europe and east Asia. The smooth line was created using an approximately fifty-year Gaussian filter. Recent instrumental data for Northern Hemisphere summer temperature anomalies (over land and ocean) are also plotted (thick line). The instrumental record is probably biased high in the mid-nineteenth century, because of exposures differing from current techniques (IPCC, 1995, Fig. 3.20).

sustained cooler period from around 1500–1800. But ice-core data from other parts of the world show a more varied picture (see Fig. 4.21).

The most obvious conclusion is that there was not a monotonously cold period in the sixteenth to nineteenth centuries. Certain intervals were, however, colder than others. Only a few short cool episodes appear to have been synchronous on the hemisphere and global scale. These are the decades 1590s to 1610s, the 1690s to 1710s and 1800s, and the 1880s to 1900s. Synchronous warm periods are less evident with the 1650s, 1730s and 1820s being the most striking. The geographical extent of climatic anomalies lacks synchronicity: the coldest episodes in one region often not being coincident with those in other regions. Roger Bradley and Phil Jones, who edited a comprehensive study of the climate since 1500 (see Further Reading), have produced a diagram which shows some evidence of cooling in summer temperatures across the northern hemisphere (Fig. 4.22). Their overall conclusion is, however, that the term 'Little Ice Age' should be used with caution.

4.10   THE TWENTIETH CENTURY WARMING

The evidence of global warming since the late nineteenth century draws principally on instrumental records. The reliability of these, and what it means for interpreting the causes of the warming are analysed in Chapters 6 and 8. The best estimate of the warming in the twentieth century is that annual global surface temperatures have risen by 0.62 °C, with the rise being slightly greater in the southern hemisphere than in the northern hemisphere

(Fig. 4.23(a), (b) and (c)). The warmest years in the record have all fallen in the 1990s. The greatest warming this century was during the two 20-year periods 1925–1944 and 1978–1997. Over these periods global temperatures rose by 0.37 °C and 0.32 °C respectively. In between there was a slight cooling, which was more marked in the northern hemisphere (Fig. 4.23(b)). In the southern hemisphere the warming trend has been

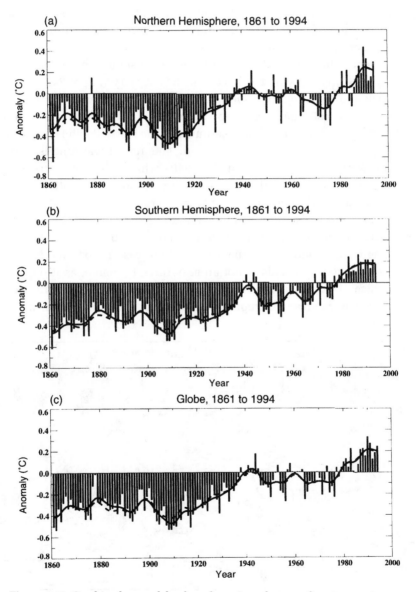

Figure 4.23 Combined annual land–surface air and sea-surface temperature anomalies (°C) 1861–1994, relative to 1961–90 (bars and solid smoothed curves): **(a)** Northern Hemisphere; **(b)** Southern Hemisphere; **(c)** Globe. The dashed smoothed curves are corresponding results from IPCC (1992), adjusted to be relative to 1961–90 (IPCC, 1995, Fig. 3.3).

more gradual and continuous (Fig. 4.23(c)). The twentieth century warming has been accompanied by a decrease in those areas affected by exceptionally cool temperatures and, to a lesser extent, increases affected by exceptionally warm temperatures. In recent decades night minimum temperatures have increased more rapidly than in day maximum temperatures. Over the period 1950 to 1993 the diurnal temperature range has decreased by 0.08 °C per decade.

Another feature of the warming this century is the regional patterns. In recent decades, the most pronounced warming has been over much of the northern continents and a marked cooling in the north-west Atlantic, with a lesser cooling across the central northern Pacific. These changes show up most clearly in winter – December to February (Fig. 4.24). On a seasonal basis the warming over the mid-latitude Northern Hemisphere continents has been restricted largely to winter and spring while the cooling of the north-west Atlantic and mid-latitudes of the North Pacific has partially compensated the overall warming trend. This recent warming is in contrast to the warming period around the 1940s which was concentrated at higher latitudes of the northern hemisphere. Also, the southern hemisphere warmed less during this earlier warm period.

The recent rise in global temperature is only part of the current concern about our changing climate. In terms of the economic impact the most damaging consequence would be a marked increase in extreme weather events (see Section 5.8). This is not an easy thing to define, as it depends on the choice of extremes and reliable measurements of such events over a sufficiently wide area to be representative of global trends. One measure that has

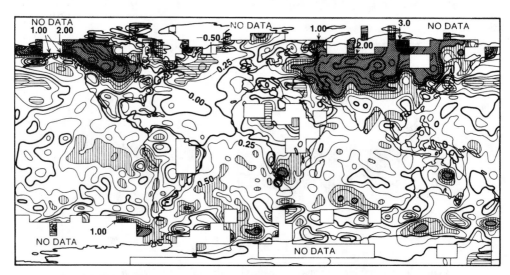

Figure 4.24 Changes in winter (December to February) temperature during the 1980s, compared with the average figures for 1951–80 showing that the most marked warming occurred at high latitudes, but that there were areas of cooling notably in the vicinity of southern Greenland (IPCC, 1992, Fig. C.5(b)).

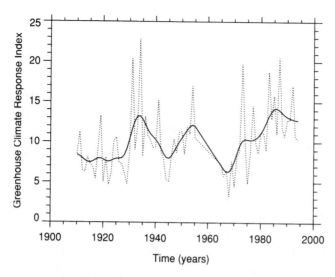

Figure 4.25 A climate index for the USA of certain weather extremes which shows an increase since the 1970s which is consistent with predictions of global warming (IPCC, 1995, Fig. 8.8).

been produced, estimates what proportion of the total area of the contermi-nous USA experienced extremes in either the top or bottom 10% of a variety of climatic indicators in any given year. Known as the US Greenhouse Climate Response Index (GCRI), this methodology selected indicators on the basis of the concerns of the public and policy-makers. It covered increases in temperature, cold season (October to April) precipitation, severe summertime (May to September) drought, proportion of total precipitation derived from extreme one-day events, and decreases in day-to-day tempera-ture variations. This index shows peaks in the 1930s and 1950s associated with the widespread droughts in these decades and the marked rise since the mid-1970s which may be linked to a shift in global circulation (Fig. 4.25).

More generally, climatologists have concluded that there is no consistent trend in interannual temperature variability in recent decades, and no con-sistent pattern for rainfall variability. The same story emerges in respect of intense rainfall and extratropical cyclones. Tropical cyclones appear to have declined in the North Atlantic but elsewhere the observations are not sufficiently reliable to draw any conclusions. Overall, there is no clear evi-dence that extreme weather events, or climate variability, has increased, in a global sense, during the twentieth century.

4.11   CONCLUDING OBSERVATIONS

The clear message is there is no shortage of evidence of climate changes on every timescale. The scale of these changes has clearly had major impacts on the world around us. But, while there is a lot of evidence, it falls far short of providing a complete picture of the size of changes and when and where they took place. This limitation applies to even recent instrumental observations and becomes increasingly the case as we go back in time as the uncertainties about the Little Ice Age and the medieval climatic optimum demonstrate. The principal features of the recent ice ages and longer term fluctuations over the geological timescale stand out, but almost in every period there remain more questions than answers.

These gaps assume greater importance in exploring the consequences of climate variability and climate change. In many instances, from mass extinctions in the distant past to assessing the implications of recent weather-related economic and social upheavals, it is not certain what part fluctuations in the climate have played in events. So, whenever climatic factors are cited as the cause of other changes, these claims need to be examined closely. The first stage must be to obtain independent evidence of the climate changing in the manner proposed. Often the available information is either too sparse and ambiguous to draw categoric conclusions, or there is a degree of circularity in the analysis. Where these objections are justified, the only approach has to be to accept that until new evidence based on improved measurements becomes available, no conclusion can be reached. Where the evidence is more convincing the second stage is to demonstrate the inferred changes could have had the impact proposed. To do this we must explore the examples of where the evidence of climatic change can be linked with other significant consequences, to establish which aspects of the climate really matter.

QUESTIONS

1   It has been suggested that one explanation for the possible occurrence of the 'Snowball Earth' (see Section 4.1) was that the tilt of the Earth's axis of rotation was much greater at the time. Why would this alter the temperature distribution between the equator and the poles and how could the axis be shifted into its current position?

2   If the area of the world's oceans is 360 million square kilometres, calculate the amount the sea level should have dropped if the amount of additional ice locked up in the ice sheets during the last glacial maximum was sixty million cubic kilometres. To the extent that the figure you calculate is greater than that given in Section 4.4, explain what other physical effects could explain this difference.

3   Many of the sources of evidence of climate change during the last 10,000 years come from only a limited number of places around the world. Explain why this is a severe limitation in trying to establish how the global climate has changed in recent millennia. The answer to this question should concentrate on why climate change might follow a different course around the world (e.g. between the tropics and higher latitudes; between continents

and oceans; and, between the northern hemisphere and the southern hemisphere).

FURTHER READING

A complete reference list is available at the end of the book but the following is a selection of the best books or articles to follow up particular topics within this chapter. Full details of each reference are to be found in the Bibliography.

Benton (1995). An excellent review article on current thinking about how life has diversified on Earth and the role played by mass extinctions in this process.

Bradley and Jones (1995). A series of papers by many of the leading lights in the study of climate change over the last 1,000 years, which provides many valuable insights into the challenges of building up a reliable picture of past climates.

Dawson (1992). This textbook on the geology and climate of the last ice age provides an accessible and comprehensive description of what is known about the global environment between 125,000 and 10,000 years ago. Its only limitation is that it does not include the latest work on the evidence of sudden changes in the climate which would make it even more valuable: a new edition is eagerly awaited.

Frakes et al. (1992). A good presentation of many aspects of climatic change on geological timescales, together with some interesting theories on the periodic nature of these long-term fluctuations.

Grove (1988). A scholarly analysis of the evidence of the Little Ice Age, which concentrates principally on the fluctuations in the extent of glaciers in the mountain ranges around the world.

Lamb (1972) and (1977). This two-volume classic work on all aspects of climate change includes comprehensive information on the early studies of the evidence of past climatic ups and downs.

Lamb (1995). A fascinating analysis of the historical impact of climatic fluctuations throughout recorded history.

CHAPTER FIVE

CONSEQUENCES OF CLIMATE CHANGE

In nature there are neither rewards nor punishments – there are consequences.

Robert G. Ingersoll, 1833–1899

Identifiable consequences of climate variability and climate change provide a measure of the significance of climatic events in our lives and the world around us. In many disciplines failure to understand how a changing climate may have influenced outcomes can lead to partial or inaccurate interpretation of past events. So identifying the most important consequences provides a checklist of the issues which require more research and are central to predicting what future changes may occur and what their impact could be.

The analysis of the consequences of past climate change falls naturally into the two areas identified in Chapter 4. First, there is the long-term and often dramatic fluctuations that occurred before the start of the Holocene, some 10,000 years ago. Then there is the Holocene with its relatively stable climate. This stability means the consequences of climatic shifts are intertwined with other events and so the central issue is whether they have played a significant part in human economic and social development. This separation does not mean there are no periods of great climatic stability before the last ice age. Indeed, as implied in Section 4.2 there may well have been vast periods of geological time when the climate was far more benign than in recent millennia. Here, however, the distinction helps to distinguish between interpreting the impact of long-term climate change on many earth sciences, and the more immediate issues of understanding how current climatic fluctuations may now influence our lives.

5.1   GEOLOGICAL CONSEQUENCES

In examining the geological record, there is the fundamental issue of whether it is possible to unravel the climatic factors from other causes of geological change. Clearly, climatic events like the waxing and waning of ice sheets, the desiccation of continental interiors or the drying up of oceans had major impacts on the geology of the large parts of the Earth. The real question is were they 'cause' or 'effect'? If they were the result of more basic processes in the Earth's geological history linked principally to plate tectonics and changed levels of volcanism then their import in understanding climate change is less profound than if they played a significant part in driving the pace and direction of the underlying geological changes. This boils down to two issues. First, do shifts in the climate exert a significant feedback on tectonic activity (e.g. volcanism) by, say, the changing load on the Earth's crust due to the build-up and collapse of polar ice sheets, or changing sea levels? Secondly, if the climate can exert influence on tectonic activity, is it affected by extraterrestrial effects (e.g. the Earth's orbital parameters, fluctuations in the Sun's output, or even the motion of the Solar System through the Galaxy)?

As yet, there is no agreement on whether climate change exerts a significant effect on tectonic activity. There is, however, sufficient evidence to suggest that there are various ways in which the climate might create, or help relieve, the stresses in the Earth's crust. For instance, during the last ice age the eruption of volcanoes in the Mediterranean region appear to have occurred more often at times of rapid sea level change. This supports the proposition that changes in the loading of the Earth's crust by the build-up and collapse of ice sheets over northern Europe did influence the level of volcanic activity in the region. How this fed back into climate change at the time is not yet clear.

A more subtle global effect of climate change occurs in the form of changes in the length of the day. If the atmosphere circulates more rapidly, the principle of the conservation of angular momentum requires that the Earth should slow down an ever so tiny amount to compensate for this atmospheric acceleration. Measurements in recent decades have shown that when climate warms up, say, as a result of a major ENSO warming event (see Section 3.7), the equatorial winds speed up and the Earth slows down. In the case of the 1997–8 event the length of the day increased by about 0.4 ms. This suggests that during sudden changes of the climate the Earth could speed up or slow down significant amounts, which in turn could set up stresses and strains in the crust which might lead to increased volcanic activity and, hence, to an additional perturbation of the climate.

As for the extraterrestrial influences on the climate, these are the subject of intense speculation (see Chapter 8). The most obvious example of these influences is the variation of Earth's orbital parameters that is seen as a major factor in ice-age dynamics. Given the rather direct nature of the proposed link between ice sheet dynamics and volcanic activity (see earlier), this is another factor that needs to be included in a comprehensive theory of climate change. As for other extraterrestrial influences on the climate, their impact is somewhat less well established, but one of the consequences of shifts in the climate is to alter tectonic activity, then the possibility of extraterrestrial influences playing slightly greater part in the process becomes yet one more factor to consider.

At the practical level, it is clear from the geological record discussed in Chapter 4 that climatic factors are part of immense changes. Reserves of coal, natural gas and oil are the product of a series of events which include major changes in the climate. For example, the sequence of a highly productive warm shallow sea laying down large quantities of organic matter followed by a period of desiccation which dried out the sea and capped the organic matter with an impermeable layer of evaporites is a classic formula for creating oil and gas fields. The fact that these processes may also have been driven by other forces, such as tectonic uplift, which isolated the sea and accelerated the rate of desiccation, does not alter the fact that the climate played an important part in the outcome. This means that understanding the geological record requires a knowledge of how both the global and regional climate behaved at the time the strata were laid down. So, it is not surprising that oil companies are among the organisations which are funding research into how the oceans of long ago deposited organic material to inform their search for undiscovered oil fields.

The same principle applies to interpreting features of the landscape. While many of the consequences of glaciation relate only to recent Quaternary events, the analysis of both glacial and other features requires a balanced approach to what can be attributed to climate change and what is caused by other processes.

5.2   FLORA AND FAUNA

The impact on flora and fauna is another area already covered in presenting the evidence of past climatic fluctuations. There are, however, certain features of using evidence of the distribution of flora and fauna in the past which need to take account of how plants and animals evolved to meet the challenge of a changing climate. So it may not be safe to assume that animals which appear to have required a warm climate (e.g. dinosaurs) were not capable of living in cooler parts of the world which exhibited a marked seasonal cycle. For instance, skeletal remains indicate that dinosaurs lived on the North Slope of Alaska around 75 Ma. At the same time this region was at a high latitude and experienced a mild-to-cold climate. The dinosaurs lived in a deltaic environment among forests of deciduous conifers and broad-leaved trees, and probably remained there all year round, coping with long winter darkness and low temperatures. While the heated debate as to whether dinosaurs were warm-blooded remains unresolved, it is clear these particular species had adapted to a challenging climate.

Similarly, the assumption that fossilised relatives of current warmth-loving plants (e.g. cycads, see later) could only have lived in year-round temperate or warm climates may underestimate the adaptability of plants long ago. The development of the plant kingdom (Fig. 5.1), and in particular vascular plants (plants having a vascular system for conducting water and food solutions) provides a good example of the interaction with climate change. These plants fall into three groups:

    i) pteridophytes (ferns, horsetails, clubmosses);
   ii) gymnosperms (conifers, cycads); and
  iii) angiosperms (flowering plants).

The pteridophytes (spore-bearing plants) were the first to appear, emerging from the protective environment of the oceans some 400 Ma. Then came the gymnosperms; woody plants bearing cones with naked seeds (i.e. not enclosed in an ovary). In this group, pollen grains from the male cones are carried by the wind on to ovules (unfertilised seeds) produced on the scales of female cones. The most recent group are the angiosperm or flowering plants, which have their ovules enclosed in a protective structure – the ovary. This usually extends upwards to form the style and stigma. Pollen

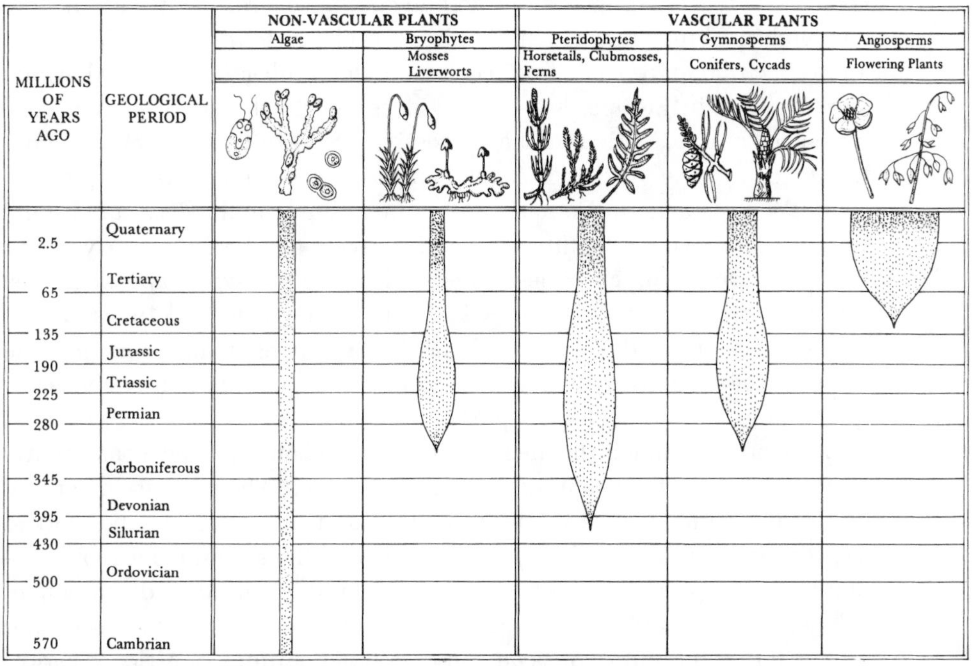

MILLIONS OF YEARS AGO	GEOLOGICAL PERIOD	NON-VASCULAR PLANTS		VASCULAR PLANTS		
		Algae	Bryophytes / Mosses Liverworts	Pteridophytes / Horsetails, Clubmosses, Ferns	Gymnosperms / Conifers, Cycads	Angiosperms / Flowering Plants
2.5	Quaternary					
65	Tertiary					
135	Cretaceous					
190	Jurassic					
225	Triassic					
280	Permian					
345	Carboniferous					
395	Devonian					
430	Silurian					
500	Ordovician					
570	Cambrian					

Figure 5.1 How the plant kingdom has developed over the geological timescale (from *100 Families of Flowering Plants*, Cambridge University Press, Hickey & King, 1988, Fig. 1).

grains landing on the stigma can only reach the ovules by forming a tube which grows down through the style and into the ovary.

Since their emergence in the early Cretaceous (around 120 Ma) flowering plants have overtaken conifers and cycads to become the dominant form of land plants and now number some 250,000 species. As such they provide the best example of how slow variations of the climate have exerted powerful controls over the development and distribution of species. For much of the last 100 million years, as the continents drifted into their current positions, the warm climate of the late Cretaceous and early Tertiary may have been a crucial factor in allowing flowering plants to diversify and increasingly monopolise terrestrial vegetation – a dominance they continue to hold to this day. By the Eocene, some 50 Ma, tropical and sub-tropical plants had extended their range as far north as western Europe, and the Pacific Northwest of the United States. The subsequent cooling trend pushed this range back somewhat.

The dramatic changes in extent of flowering plants is more easily discerned with the onset of the Pleistocene and the periodic expansion of the ice sheets in the northern hemisphere. In North America the plants were able to migrate north and south with the shifting climatic zones. So, many of the consequences for plants were limited, and can only be detected in pollen analysis (Fig. 5.2), which show the waxing and waning of many species with rising and falling temperatures. In Europe, the barrier of the Alps posed a greater challenge to migration north and south. As a result, species that are common in China and the United States, such as *Actinidia* (the Kiwi fruit family), *Liriodendron* (Tulip Tree) and *Liquidamber* (sweet gum) which were present in Europe, but could not escape to the south of the Alps, were driven to extinction by the successive waves of ice. This process continues to this day with the extent of various species adapting rapidly to shorter fluctuations. In Denmark the population of holly (*Ilex aquifolium*) was largely wiped out by three successive severe winters in the early 1940s, while the range of the Nine-banded Armadillo (*Dasypus novemcinctus*) spread north through Texas and Oklahoma in the 1940s and 1950s only to retreat from the colder winters of the 1960s. Changes like these provide evidence of both past climate change and the possible consequences of future global warming (see Chapter 10).

Even more striking is the theory that climatic change played a particularly important part in the evolution of modern humans. The stepwise deterioration of the climate since the mid-Pliocene and, in particular, the marked cooling event around 2.5 Ma, perhaps provided the stimulus for rapid evolution of our species. The increasing aridity of this period, and subsequent cold episode, would have favoured cold and drought-resistant taxa and provided the conditions for savannah-dwelling hominids in Africa to evolve more rapidly. The drier, more open landscape with more pronounced seasons led to the need to forage further and longer and provided the spur for early hominid anatomy and physiology to be exposed to novel selective

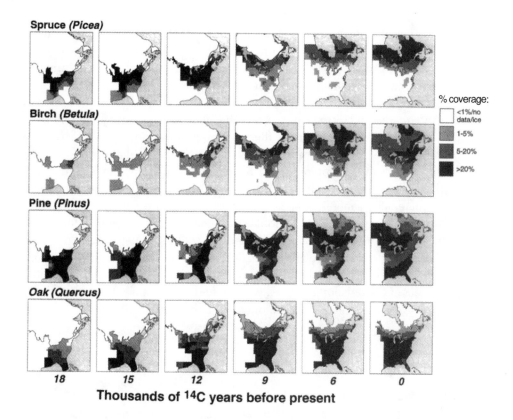

Spruce (Picea)

Birch (Betula)

% coverage:

<1%/no data/ice

1-5%

5-20%

>20%

Pine (Pinus)

Oak (Quercus)

18   15   12   9   6   0

Thousands of ^{14}C years before present

Figure 5.2 Postglacial changes in the distribution and abundance of some major tree types in eastern North America, based on pollen analysis data (IPCC, 1995, Fig. 9.3).

pressures. In particular, the development of bipedalism and relative hairlessness appear to have been the response to these challenges. So our very existence may be a consequence of the closing of the Isthmus of Panama (see Section 4.1), and the ice ages which started around 2.5 Ma.

The emergence of modern humans (*Homo sapiens sapiens*) during the last ice age, and the disappearance of Neanderthals (*Homo neanderthalensis*), around 35,000 years ago, may also be the consequence of the dramatic climatic fluctuations which occurred between 50,000 and 15,000 years ago (see Section 4.4). Like so many aspects of evolutionary theory, this is an intellectual snakepit. Put at its simplest level this is an area where there are many more theories than pieces of evidence. From what little we know it is apparent that intelligent tool-making hominids have inhabited Europe for about half a million years. But modern humans emerged from Africa, appearing in the Middle East around 100,000 years ago and lived alongside Neanderthals for many tens of thousands of years, although the DNA evidence suggests there was no interbreeding. The latter were physically well-endowed to survive the capricious climate of the last ice age, having an immensely powerful physique adapted for endurance and prolonged locomotion over irregular terrain. Nevertheless, by the time the ice age reached its climax they had disappeared, and it was modern humans, with their more advanced tool-making skills, who were able to exploit the more benign post-glacial conditions.

An equally contentious area is whether there is a climatic component in the emergence of agriculture at the beginning of the Holocene. Clearly, the stability of the climate over the last 10,000 years has made agriculture a viable proposition. Moreover, we have no way of knowing whether any earlier attempts to develop agriculture were snuffed out by sudden changes in climate. The general view is, however, that the pressures of population growth and improvements in tool technology were the driving forces for establishing agriculture. Population growth may have been the product of an improving climate. We must be careful not to press the climatic argument too far, however, especially as there is no supporting evidence beyond the basic fact of the development of a warmer and more stable climate following the ice age coincided with these developments.

5.3   MASS EXTINCTIONS

Changes in fauna over geological timescale have been massive, but climate change is only part of the story. In particular, its role in mass extinctions is a fertile area of scientific debate. These arguments centre on not only the part played by climate but also on what happened during these upheavals, how quickly they developed, and even just how many of them there were. Much depends on the quality of the fossil record, and its ability to document the pattern of the history of life. While some palaeontologists take the view that the record is sufficient to draw conclusions, others argue that before around 300 Ma the record is inadequate because less fossils have survived and less work has been done to find those that remain.

Over the last 600 My, even allowing for the limitations of the early part of the record, a broadly consistent picture emerges. This shows that the diversity of all types of marine and continental life including microbes, algae, fungi, protists, plants and animals, increased exponentially since the end of the Precambrian, and now totals between five and fifty million species. This diversification was interrupted by mass extinctions, the largest of these included the 'Big Five' (see Section 4.1), at 440, 365, 255, 210 and 65 Ma. Pride of place goes to the Late Permian extinction (255 Ma) in which just over 60% of all families of species were wiped out, with possibly as many as 96% of all species disappearing, with the mortality being higher on land than in the oceans. A further mass extinction in the early Cambrian at 520 Ma could be added on to this list, although the number of species involved and uncertainties about how rapidly it occurred mean it has to be treated with caution. Furthermore, a number of more minor, but still significant extinctions, have been identified in the fossil record, the last of which was in the late Eocene around 38 Ma, and coincided with a marked cooling in the climate (see Section 4.1). The identification of these less significant events depend in part on the improved fossil record as we get close to the present. Nevertheless, the picture is of a continuum of events with increasingly frequent minor extinctions being responsible for the elimination of many species. What is not clear is whether the 'Big Five' fit in to this continuum, or are in some way fundamentally different. But, overall, species are at low risk of extinction most of the time, and this condition of relative stability is punctuated at rare intervals by a vastly higher risk of extinction.

Various attempts have been made to establish a pattern in the timing of the major extinctions. Recent analysis (see Further Reading) does not support these efforts, and concludes that the seven identifiable mass extinctions in the last 250 Ma, were spaced twenty to sixty million years apart and there was no periodicity in their occurrence. So, to the extent that we might expect climatic change on this timescale to be periodic (see Section 8.9), this conclusion reduces the possibility that mass extinctions were the product of climatic events. But, if random cataclysms (e.g. collisions with asteroids or comets, sudden huge bouts of volcanic activity, or simply the capacity of non-linear responses (see Section 8.1) of the global ecosystem to produce catastrophic circumstances) precipitated these evolutionary crises, climate change was almost certainly part of the consequent upheaval.

Whatever the role of the climate in mass extinctions the important point to note is how throughout geological time the diversity of life has increased. While extinctions and other environmental changes have resulted in major setbacks, the upward trend is clear. This means that, in adapting to these changes, many lifeforms evolved in conditions which were very different to those in which their descendants are now living. So we have to be careful when drawing conclusions about past climates on the basis of geographical distribution of certain types of flora. The second point is how this adaptability will come into play in responding to future climate change. As many species have developed genetic defences to past changes, some of them will be equipped to meet future challenges. What we do not know is which species will survive, nor which genetic defences will provide the best protection against both natural variations and the consequences of human activities.

5.4 GLACIERS, ICE CAPS, ICE SHEETS AND SEA LEVELS

The changing volume of freshwater locked up as land ice around the world has various consequences. Most obviously it affects worldwide changes in sea level (often termed *eustasy*). The collapse of the North American and Fennoscandian ice sheets following the last ice age rapidly produced a rise in sea level. This came to a virtual halt about two thousand years ago (Fig. 5.3). The precise changes in sea level between around 5,000 and

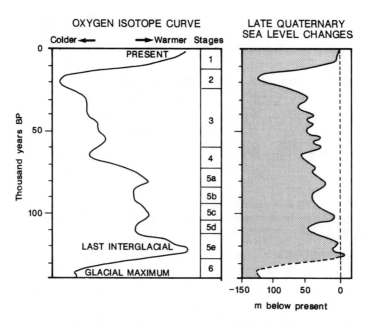

Figure 5.3 Changes in ocean sediment oxygen isotope content and in sea level since the last interglacial. Analysis for various parts of the world show somewhat different detailed changes especially in the last few thousand years, with some evidence of higher levels around 5000 BP (Van Andel, 1994, Fig. 4.12).

2,000 years ago remain the subject of dispute, and this uncertainty is reflected in the alternative curves shown in Fig. 5.3. What is clear is that over the last two millennia the sea level has varied within a range of no more than a few tens of centimetres.

In terms of the last century, the principal factors contributing to the sea level rise are the melting of glaciers and ice caps around the world, but not including the ice sheets of Antarctica and Greenland, and thermal expansion of the oceans. Oceanic thermal expansion is a consequence of global warming (see Section 4.10). Using computer models it is estimated that the expansion due to this warming falls in the range 2–6 cm with a median value of about 4 cm. The contribution from the general recession of glaciers and ice caps (Fig. 5.4) may have amounted to about 4 cm. Other potential contributors include changes in the mass of the Antarctic and Greenland ice sheets, extraction of ground water under the continents for agricultural purposes, and changes in the volume of inland lakes and seas. In each case the analysis is equivocal, and the overall assessment is summed up in Table 5.1.

Locally, changes in the extent of glaciers and ice caps affects the lives of those who live in their shadow. On the larger scale the waxing and waning of the ice sheets during the ice ages have left their mark on the landscape (see Sections 4.5 and 5.1). These changes also exert a continuing impact in terms of the deformation they caused to the Earth's crust. The term *isostasy* is given to the theory which proposes that where the crust is thickest it extends into the mantle beneath it. Because crustal material is less dense

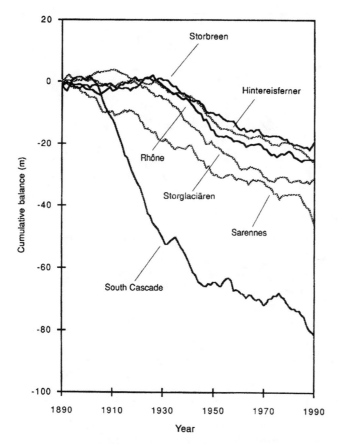

Figure 5.4 Cumulative mass balances, in metres of water equivalent, for the glaciers Hintereisferner (Austria), Rhône (Switzerland), Sarennes (France), South Cascade (United States), Storbreen (Norway) and Storglaciären (Sweden). These are among the few glaciers with long observational time series that have been extended using well-calibrated hydrometeorological models. All values are relative to 1890 (IPCC, 1995, Fig. 7.3).

TABLE 5.1 ESTIMATED CONTRIBUTIONS TO SEA LEVEL RISE OVER THE LAST 100 YEARS (IN CM)

Component contributions	Low	Middle	High
Thermal expansion	2	4	7
Glaciers/small ice caps	2	3.5	5
Greenland ice sheet	−4	0	4
Antarctic ice sheet	−14	0	14
Surface water and groundwater storage	−5	0.5	7
TOTAL	−19	8	37
Observed	**10**	**18**	**25**

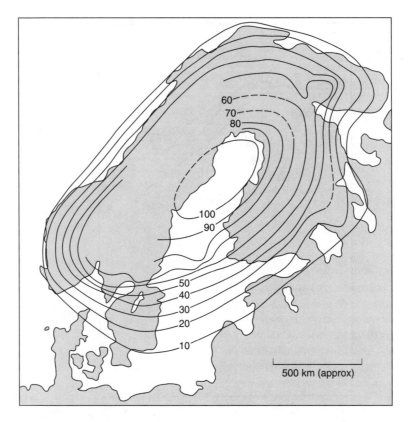

Figure 5.5 A map of the present elevation in metres of the shoreline of 5,000 years ago in Scandinavia to illustrate isostatic 'rebound' following deglaciation (*Cambridge Encyclopaedia of Earth Sciences*, Fig. 15.5).

than the mantle, it effectively forms a 'root' under mountain ranges which holds them up. The same is true where an ice sheet builds up. When the ice sheet melts away, the less dense crustal 'root' is no longer needed to hold it up, and rebound occurs. During the height of the last ice age around 18,000 years ago the additional load on areas like the Canadian Shield and the Baltic Shield caused the crust to dip downwards by as much as 700 m. This was accommodated by the creep of mantle material away from the region of application of the extra load. Melting of the northern hemisphere ice sheets caused a reverse flow of mantle material and slow uplift (*glacial rebound*) which continues today at rates up to 10 mm per year (Fig. 5.5). This means that in some coastal regions in North America and northern Europe this process is of as much importance as more general sea level rises.

5.5   THE HISTORICAL IMPACT OF CLIMATIC VARIATIONS

We now make a sea-change in our approach to the consequences of climatic fluctuations, in turning to the relative calm of the Holocene. In terms of the definitions presented in Chapter 1, this effectively means moving from issues of climate change to climate variability. While there have been significant fluctuations during the last 10,000 years, by comparison with the cataclysmic changes during and after the last ice age they amount to no more than minor variability. This makes it more difficult to interpret their impact in respect of rapid human social development. Nevertheless, the role of these fluctuations in the ups and downs of ancient civilisations is a source of enduring historical fascination.

The sudden collapse of apparently successful social structures has often been attributed to climatic factors. The extent of human activities was obviously affected by the better-documented changes in the climate. So the decline of pastoral exploitation of the Sahara plus the decline of the civilisations in the Euphrates valley at a time of increasing desiccation around 2,200 BC, and the abandonment of lakeside settlements as water levels rose in northern Europe around 1,200 BC can be attributed to sustained shifts in the climate at the time (see Section 4.7). But, the disappearance of major civilisations poses a far greater challenge. The collapse of the Harrapan culture, which thrived in the Indus Valley between 2,500 and 1,500 BC is a good example, as is the dark age that descended on the eastern Mediterranean around the end of the thirteenth century BC with eclipse of Mycenae, Ugarit and the Hittite Empire. It has been argued that prolonged drought and poor harvests are the explanation, possibly linked with the cooling in Europe that led to widespread migration of peoples.

Many other factors can be used to explain decline involving warfare, invasion and social instability. Moreover, in arguing the climatic case we must not fall into the trap of selective use of the evidence. In particular, the problem with dating events (see Section 6.5) makes it all too easy to attach too much significance to what seem like striking coincidences. Only when we get better measurements of how rainfall may have varied in these areas from, say, tree rings, together with more accurate dating of when social collapse actually occurred, will it be possible to make a more informed estimate of the role of climatic events.

The decline of the classical Mayan culture in the ninth century AD provides a good example of how the climatic case can be constructed. This civilisation reached a pinnacle in the eighth century when the population density in the Mayan lowlands, which extend over modern day Guatemala, Belize, Honduras and Mexico, was far higher than current levels. Because of the accuracy of the Mayan calendar and their propensity to erect large monuments containing detailed records of events, it is possible to date the cataclysmic decline accurately. Furthermore, recent measurements of sediments in Lake Chichancanab, in what is now Mexico, show that the period around AD 750 to 900 was the driest in the last 8,000 years. This suggests that a period of drought may have reduced the capacity of society to support an overbearing theocracy and building major monuments. This would have increased the susceptibility of many Mayan cities to revolt against these demands. But it would depend on each city's circumstances and this could explain why the cessation of records in different cities occurred at different times, with those cities on riverside location maintaining records longest.

Even more clear cut is the disappearance of the Norse colony in Greenland. The availability of records of contacts with the colony from its establishment in the late tenth century until the fourteenth century, together with data from ice cores drilled in the Greenland ice sheet (see Section 6.4.2), provide a pretty clear picture of climatic deterioration. But, while the colder weather is not in dispute, the extent to which it destroyed the colony is still the subject of debate. An alternative explanation is that the rigidity of the social structures and the reluctance to adopt Inuit technology, more appropriate to the harsh climate, may have had as much to do with the collapse as the cooling trend.

Turning to more recent history, the shifts in the climate identified over the last millennium (see Sections 4.8 to 4.10), while significant, have not exerted an overwhelming influence on the course of history. Indeed the argument among most historians is whether these changes have had any appreciable impact whatsoever. So, rather than seeking to establish the shadowy consequences of climatic shifts on history at large, it is better to examine those social and economic activities where changes are most likely to have had an impact and then ask the question as to whether they could have had wider implications. This approach also prepares the ground for considering questions of how future changes in the climate may impact on the most vulnerable sectors of our modern society.

5.6   AGRICULTURE

Of all human activities, agriculture is the most sensitive to both weather fluctuations and climate change. Furthermore, prior to the industrial revolution, in medieval Europe the purchase of food represented some 80% of the expenditure of working people. So fluctuations in food supplies and hence prices had an immediate and costly impact on the majority of the population. When famine occurred then the death rate rose. Examination of mortality statistics in England in the seventeenth and eighteenth centuries concluded that the cumulative effects of weather-induced price variations over five years was essentially zero, as all they did was alter the timing of deaths which would in any case soon have occurred.

The socially disruptive effects of high prices and food shortages is more difficult to quantify. Many historians have proposed links between such fluctuations and more widespread disorder. For instance, the French Revolution in 1789 is sometimes linked with damaging summer storms which compounded the already vulnerable food supplies and spread social unrest. But the scale of the changes in the climate at this time were not that exceptional and so placing too much emphasis on specific extreme weather events in the preceding year or two seems to overlook wider social and economic factors. To be convinced of the connection we need to find reasons why many other equally unpleasant bouts of weather did not produce similar outbursts of civil unrest. The explanation is probably that, while weather-induced price rises may have a catalytic effect in stimulating unrest, the other social and political factors, which contribute most to revolt, were not present.

This cocktail of factors makes it equally hard to identify the overall impact of periods of climatic deterioration on agricultural societies. As noted in Section 4.9, the Little Ice Age, rather than being a sustained period of lower temperatures, was, for much of the time, not appreciably colder than the twentieth century but was punctuated by markedly colder decades which hit different parts of the world at different times. In Europe the greatest agricultural disasters were in the 1310s, the 1590s and the 1690s. Each, in its own way, provides useful insights into the major short-term disruption.

Although the extent of the Medieval Climatic Optimum (see Section 4.8) is uncertain, there is considerable evidence that by the late thirteenth century the climate in Northwest Europe had become stormier and an analysis of historical records of winter weather in Europe during the fourteenth century suggests the first quarter of the century featured an exceptional number of cold winters. But it is the run of extraordinary cold wet summers from 1314 to 1317 and their associated harvest failures that comes ringing down through recorded history as the greatest weather-related disasters ever to hit Europe. From Scotland to northern Italy, from the Pyrenees to Russia there are an unparalleled number of reports of failed harvests, starvation and pestilence. Grain prices rose to levels unsurpassed in the subsequent 150 years. Where detailed records exist, mortality rose dramatically with over 10% of the population of the rich Flemish town of Ypres dying in the summer of 1316. But, in spite of the awful impact at the time the lasting consequence of this series of bad summers was small. What they demonstrated was the vulnerability of a population whose numbers were at the limit of what could be provided for by the agriculture of the time. The arrival of the Black Death in 1348 and the almost constant warfare throughout the continent for much of the century reduced the population to a level where the vagaries of the weather had less impact on social history.

The link between the climate, agriculture and the ability to feed a growing population re-emerged as a major issue at the end of the sixteenth century. As noted in Section 4.9, this was a period of more frequent cold winters and cool wet summers in northern Europe. In England, poor harvests and rising grain prices meant the standard of living of the working people fell to the lowest level recorded in the last seven centuries (Fig. 5.6). The growing social unrest led to panic legislation. Parliament passed a Great Act codifying a mass of sectored legislation and local experiments in poverty relief. It also restored many of the restrictions on enclosures of common land and the conversion of arable land to pasture which had been repealed only four years before.

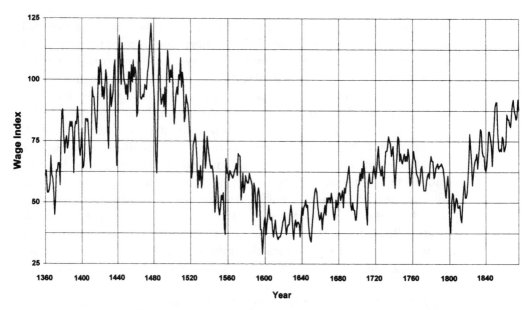

Figure 5.6 The index of the purchasing power of builders' wages in England over six centuries (Burroughs, 1997, Fig. 2.5).

Although agriculture was always stretched to meet the needs of the population in Europe during the seventeenth century, the exceptional cold of the 1690s had serious consequences. The scale of impact varied from place to place, and this provides insights into the vulnerability of different communities. France, with its high population levels, was hit first, with two poor harvests in 1693 precipitating the worst famine since the early Middle Ages. England, by comparison, escaped relatively unscathed, but the continued cold hit Scandinavia particularly hard. In Finland the famine of 1697 is estimated to have killed a third of the population. But, possibly the most lasting effect was in Scotland. Here, between 1693 and 1700, the harvests, principally oats, failed in seven years out of eight in upland parishes. Death rates rose to a third to two-thirds of the population in many of these parishes, exceeding the figures recorded during the Black Death. The economic consequences of these awful years probably, more than anything, made the union with England in 1707 inevitable.

In other parts of the world, the agricultural consequences of climate fluctuations have more often been the result of drought. There are many examples of how sustained drought has destroyed apparently thriving farming communities, but it is only where this leads to a radical change in managing subsequent activities that it can be said to have had real consequences. A good example of this response is how the United States Government rose to the challenge of the Dust Bowl years of the 1930s. Ever since settlers had introduced arable farming to large areas of the Great Plains, periodic series of hot, dry summers had destroyed crops and driven many farmers from the land. Both in the 1890s and the 1910s many areas experienced widespread depopulation as crops failed.

In 1934 and 1936 the average wheat yields across the Great Plains fell by about 29% compared with the trend. In the hardest hit regions of Kansas and Oklahoma there was massive outward migration with more than half the population leaving. But the Democratic Government recognised that without central action the problems of recurrent drought would continue. It concluded that much of the agriculture on the Great Plains was not appropriate to such an arid region. So marginal land was purchased and retired from cultivation and seeded with grass. This was combined with educational programmes for farmers to plant trees for shelter-belts, grow crops better suited to drought conditions and introduce conservation methods (e.g. contour ploughing, water conservation and strip ploughing to allow part of the land to lie fallow). Although the pressures of the Second World War led to some marginal land being pressed back into service in the wetter 1940s, subsequent droughts have built up the acceptance of the need for state and federal laws to protect the land from over-exploitation.

In recent decades the problems of drought have been most extreme in the subtropics, notably in sub-Saharan Africa (the Sahel). The drought in the Sahel started in earnest in 1968 and reached a nadir in 1972. It then abated, only to return in the late 1970s and worsened during the 1980s before easing off in the 1990s (Fig. 5.7). Its most dramatic social consequences were during the initial phase when over 100,000 people died and the pastoral economy of the more arid regions was effectively destroyed. These awful events had a dramatic impact on thinking in respect of both the provision of international aid and the consequences of both the natural variability of the climate and impact of human activities on arid regions (see Section 8.9). The famine in Ethiopia in 1984 reinforced these concerns.

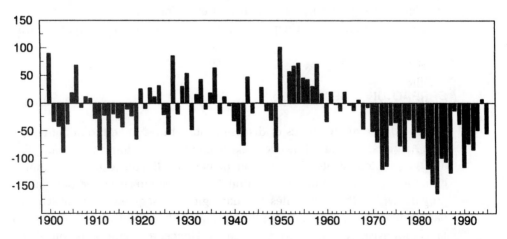

Figure 5.7 Standardised precipitation anomaly in the Sahel Region showing the sharp reduction in rainfall from the late 1950s onwards (with permission of UK Meteorological Office) (Burroughs, 1997, Fig. 1.3).

Analysis of the events in the Sahel, together with lengthy studies of the variation of the Indian monsoon, has led to the conclusion that tropical sea surface temperatures are a major factor in rainfall variations in the sub-tropics (see Further Reading). In particular, the ENSO (see Section 3.6) is a useful predictor of wet and dry years – poor monsoons in India are correlated with warm El Niño events in the eastern Pacific. So, the quasi-cyclic fluctuations of the climate, owing to the ENSO, have widespread agricultural consequences. But, understanding their behaviour offers the prospect of long-range seasonal forecasts which may enable farmers to make decisions about planting the right crops for dry or wet seasons. Already countries as far apart as Brazil, Peru and Zimbabwe have used these forecasts to plan agricultural activities.

Although the prospect of improved forecasting may lead to some advances in the ability of agriculture to respond to climate variability, the general message is that we remain as vulnerable as ever to extreme weather events. This means that if the climate should become more variable or undergo some more marked change, the implications for food production around the world are bleak. In essence, any change will be a bad thing, especially as the growing population of the world makes ever greater demands for increased productivity.

This gloomy conclusion, together with recent painful lessons of the consequences of prolonged extreme weather disrupting agriculture and rural communities exerts a major influence on current farming policy in many countries. It is recognised that without some government involvement, extreme weather events exaggerate other fluctuations in supply and demand and do great damage to both farming interests and the community at large. So, possibly the most important consequence of future changes in the climate for the agricultural sector is to identify the right level of government intervention in managing the markets. The ability to extract the right message from past agricultural experience in planning for future climate variability (see Chapter 10) will be an important part of this process.

5.7   SPREAD OF DISEASES

The incidence of epidemics of disease is sometimes linked to climatic factors. Any analysis must, however, take account of variations in food supplies and population levels. So a period of benign climate (e.g. the medieval climatic optimum in Europe – Section 4.8) can lead to sustained population growth, which then magnifies the damaging consequences of subsequent bad weather. Reduced food supplies at times of historically high population levels increases vulnerability to diseases. Hence, any interpretation of the role of climate variability in famine and social decline has to take account of demographic trends.

There are additional intrinsic reasons for the coincidence between major pandemics and periods of serious social disruption. These are part of the very nature of many of the diseases which afflict the human race. The symbiotic relationship between a disease (e.g. bubonic plague, influenza or typhus), and its animal and human hosts means potentially more virulent mutations in the disease organism may remain quiescent because the local animal or human populations have some degree of immunity. When, however, some major upheaval occurs owing to, say, drought, earthquakes or floods, populations are dispersed, mixing with others who have less immunity, and a new plague may emerge with horrifying rapidity.

Examples of this chain of events which may have had a climatic component include the Justinian Plague of the sixth century, the Black Death in the fourteenth century and the global cholera pandemic of the 1830s. Each of these cases is worth examining as it provides insights into the complexity of the processes involved. The epidemic that struck Constantinople in 542, in the reign of Emperor Justinian, was the first reliably identified instance of bubonic plague. While historical sources claim it came from Ethiopia, its origin may be linked to the sudden cooling associated with the 'mystery cloud' of 536 (see Section 8.10). What is clear is the consequent crop failures from Rome to China which led to widespread famine, and against the background of growing demographic pressures, this provided the right combination of social pressures ideal for the emergence of a pandemic. There is no way of quantifying the contribution of climatic disturbance to these events. Nevertheless, over the next century or so, the population of Europe halved and the more heavily populated Mediterranean region may have suffered more grievously, thereby ensuring this part of the world would be condemned to centuries of the Dark Ages.

The analysis of the origins of the Black Death involves the same complex mixture. The combination of demographic pressures, bad weather and crop failures was already a feature of northern Europe in the early fourteenth century (see Section 5.6). But, as noted in Section 4.8, there is no clear evidence of a coherent climatic trend across the northern hemisphere at the time. So, given that the plague appears to have originated in China, a more local explanation may need to be invoked. China was devastated by horrendous floods in 1332, which reportedly killed several million people, disrupted large parts of the country and caused substantial movements of wildlife, including rats, in which bubonic plague is endemic. This disruption may have triggered the emergence of the Black Death but, given that China is often wracked by devastating summer floods, once this virulent new version of bubonic plague had come into being, it was only a matter of time before it made its presence felt. So, while a climatic event may have set the ball rolling, it was only a tiny but essential component of the subsequent cataclysmic pandemic.

The cholera pandemic is an even more complicated story. Cholera appears to have been unknown outside Bengal before 1800. In 1815 the massive eruption of Tambora (see Section 8.4) appears to have disrupted the global climate, and may have been the cause of harvest failures in Bengal. The resulting famine triggered the first cholera pandemic, which eventually reached Europe and the eastern United States in 1832. In Russia the mortality was extremely high and in New York the death rate exceeded 100 per day in the summer of 1832.

These historical examples of how climatic shifts may have triggered global pandemics are of relevance to the analysis of the potential impact of global warming (see Section 10.3). The possible spread of tropical diseases and their insect vectors to higher latitudes (e.g. malaria) is widely regarded as one of the more worrying features of current climatic change. The experience of the spread of various tropical diseases during recent ENSO events provides insight into the potential for these changes to occur. But such concern must be viewed in the context of the advances that have been made in public health in the last century. Although the impact of the wider spread of tropical diseases is the source of concern, the scale of any such epidemics is likely to be tiny compared with the events described above.

5.8   THE ECONOMIC IMPACT OF EXTREME WEATHER EVENTS

The economic impact of more frequent extreme weather (e.g. droughts, heatwaves, hurricanes and winter storms) is the source of much comment, in the context of global warming. This is a difficult issue to deal with in terms of the definitions we have been working with, as it can lead to the selective interpretation of perceived changes in climate variability. Nevertheless, despite the fact that there is, as yet, little evidence of a significant shift in the incidence of extreme events (see Section 4.10), it is widely assumed that global warming will lead to greater climate variability. An examination of the impact of past extremes is therefore part of the process of both interpreting current trends and also predicting the future impact of climate change.

The most damaging forms of weather extreme during the twentieth century have varied from one part of the world to another. In many parts of the tropics and subtropics, failure of seasonal rainfall causes the greatest hardship. The industrialised countries in the mid-latitudes experience a wider variety of damaging events. For example, in North America the combination of blizzards, floods, hurricanes, tornadoes, summer droughts and heat waves, or winter freezes can all cause immense economic damage. Europe experiences a similar mixture, although hurricanes and tornadoes are not a comparable threat, while intense extratropical depressions are a major issue.

Measuring changes in these extremes faces a number of challenges. An example of this type of analysis was shown in Fig. 4.25. This provides some evidence of an increase in the types of extreme which might be associated with global warming, but it has two serious limitations in terms of assessing potential economic impact. First, there is the question of whether the changes recorded are real or the product of growing public awareness and increasing population in areas where extreme events occur. A good example of this phenomenon is the rise in the number of tornadoes recorded in the US since the 1950s (Fig. 5.8). Clearly the number of observations has increased, but what is less certain is how much is the result of climatic developments and how much is due to more efficient reporting of smaller twisters. Other measures of the impact of tornadoes show that the number of deaths each year has declined since 1950, but this could be attributed to more effective warnings leading to people taking shelter. Perhaps more significant, the number of days each year when tornadoes have been observed has shown little increase in recent decades and the number which have caused fatalities each year has shown no rise.

The second factor, which must be addressed is whether, in focusing on events related to global warming, the index overstates the economic impact of any trend. If the observed rise in the events used in the index is matched by, say, a decline in severe winter weather and killing spring frosts then the economic consequences are greatly reduced.

A more direct measure of economic impact is trends in insurance losses. These have been rising rapidly in recent decades. But, even this increase must be treated with caution. Any analysis must take full account of changing levels of insurance cover, population densities in vulnerable areas (e.g. coastal sites), inflation and even fraudulent claims. These corrections must be combined with the fact that there is little evidence of an increase in extreme events and a few major disasters can have an excessive influence on thinking (e.g. Hurricane Andrew hitting Miami in 1992 or the storms of October 1987 and January 1990 in the UK). Only by taking full account of all the relevant factors and also a sufficiently lengthy perspective is it possible to form a balanced view about whether there is a significant climatic component in current economic developments.

Hurricane losses in the United States provides a good example of how this type of correction needs to be applied. The Gulf and East Coasts of the United States suffer on average nearly $5 billion (1995, US $) hurricane damage each year. But over 83% of this damage is caused by intense hurricanes which make up barely a fifth of the tropical cyclones which make landfall along these coasts. Moreover, a single massive storm can have a disproportionate impact on thinking. For instance, Hurricane Andrew is estimated to be the most costly storm in US history causing some $30 billion in damage. Following close on the heels of Hurricane Hugo in 1989, which caused some $6 billion-worth of damage, there was a widespread assumption that these huge losses were the product of a dramatic worsening in the climate.

What was missing in this instant reaction to such major events was a thorough analysis of the trends in the costs of hurricane-caused damage along the United States coast which took account of not only inflation, but

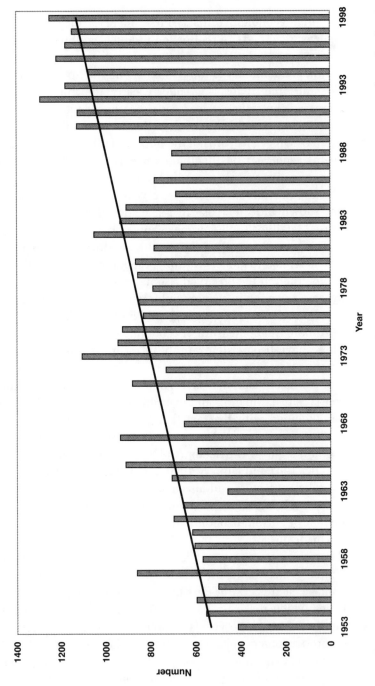

Figure 5.8 The number of tornadoes observed in the United States between 1953 and 1993 showing a strong upward trend. What is not clear is whether this trend is the result of climate change or more comprehensive observations.

also changes in coastal population and wealth, both of which have risen dramatically in recent decades in areas where hurricanes strike. When this normalisation process was conducted, it demonstrated that there has not been a rising trend in damage in recent decades. Instead there have been substantial multidecadal variations in normalised damages: the 1970s and the 1980s actually incurred less damages than the preceding few decades. Only during the early 1990s did the level of damage approach the high level of impact seen in the 1940s to 1960s. Furthermore, to put Hurricane Andrew into perspective, it is estimated that if the hurricane, which devastated Miami in 1926, were to occur now the damage would exceed $70 billion.

The implications of these comments is that, while variations in the incidence of extreme weather events represent the most immediate threat of climate change, interpretation of claims of increases in specific types of extreme require careful analysis. So we need to develop reliable up-to-date estimates of the potential damage caused by various types of extreme, and accurate measures of how these extremes will increase or decline if the climate changes in a given manner. Only then can we make the type of forecasts about the economic consequences of these changes to different parts of the world which are needed to inform policy decisions about investment to prevent or accommodate future climatic developments.

5.9   SUMMARY

The consequences of changes in the climate are, in theory, substantial. But they must not be exaggerated, by focusing only on the downside. It is essential that any analysis of the impact of climatic events accurately reflects all aspects of the argument. This need for balance works both ways. In examining the past, only a thorough appreciation of the other explanations of what may have caused changes can enable climatologists to demonstrate that climatic factors offer the key to better understanding. This means demonstrating that without the climatic component other explanations of, say, historic events are insufficient. Conversely, in considering the future, an accurate assessment must be made of how the consequences of changes in the climate compare with other threats confronting society.

The need for balance and a sense of proportion applies not only to interpreting past events but also highlights the importance of better measurements of past climates. In many instances, the weakness in the climatic explanation of past changes lies in the inability to establish precisely when and by how much the weather shifted around the time in question. This applies not only to the rise and fall of civilisations, the occurrence of famines, the changing patterns of flora and fauna, and the spread of diseases, but also to more remote questions such as the evolution of the human species and the occurrence of mass extinctions. But, in using this analysis of the past to inform our thinking about the potential consequences of future climate change, we must be sure that there is no real bias in our measures of the past. So we must now turn to the question of what can be measured, how these measurements must be analysed to get the most out of them, and what are the prospects of getting an improved picture of what really happened in the past.

QUESTIONS

1   During past changes in the climate many forms of flora and fauna were able to migrate to areas where the new climate suited them. Why could this option not be available to many species if the climate changes in the future?

2   Why could the cessation of written records by past civilisations give an exaggerated impression of social decline?

3   List the various impacts of the ENSO around the world in terms of whether they have a negative or positive impact on social and economic affairs. To what extent do these impacts tend to cancel one another out, and, if they do, are there ways in which those who lose out can be compensated by those who gain?

4   What action could insurance companies take to reduce their exposure to loss from various forms of extreme weather event and who would suffer from such action?

FURTHER READING

A complete reference list is available at the end of the book but the following is a selection of the best books or articles to follow up particular topics within this chapter. Full details of each reference are to be found in the Bibliography.

Brown, Hawkesworth and Wilson (1992). This contains a wide variety of geological information including useful discussions of palaeoclimatology and volcanoes.

Burroughs (1997). A detailed analysis of past examples of the economic and political consequences of climatic change and assesses their implications for the future.

Kates et al. (1985). An interesting selection of papers on the environmental impact of various climatic events by leading authors in their fields.

Lamb (1995). A fascinating review of the influence of climate on human history by the leading authority in the field.

Mintzer (1992). A set of papers on the challenges facing society in adapting to future climate change, drawing on the experience of recent decades.

Van Andel (1994). An intensely illuminating set of discussions of many aspects of the geological consequences of climate change.

CHAPTER SIX

THE MEASUREMENT OF CLIMATIC CHANGE

The one duty we have to history is to rewrite it.
Oscar Wilde

To be certain about how the climate has changed in the past we would need reliable measurements of the relevant parameters (e.g. temperature, rainfall, cloud cover, extent of winter snow and sea ice etc.) for representative points around the globe, for regular intervals of time going back as far as we wanted. In practice we have none of these things. Even with modern observations systems there are gaps in our knowledge of how the global climate is changing. Going back in time the problems mount. This is apparent in the evidence presented in Chapter 4. The whole process of improving the measurement of climate change is designed to fill in gaps in our knowledge of the past and establish a better foundation for the theories of why changes have occurred. The objectives of this chapter are:

- to explore briefly the most essential features of various methods of measuring past climatic conditions;
- to establish their strengths and weaknesses; and
- to investigate whether the various measurements can be combined to provide a more coherent picture of climate variability and climate change.

This involves going over some well-worn ground concerning the measurement of meteorological parameters. But, to achieve our objectives we must identify those features of measurement which limit our ability to draw conclusions about how the climate has behaved in the past and how it is currently changing.

6.1   INSTRUMENTAL OBSERVATIONS

In principle, modern instruments are capable of making most of the observations needed to study climate change. But to do so they must measure the same thing under the same conditions, wherever they are. Temperature measurement provides good examples of how difficult this basic requirement can be. While a properly calibrated thermometer can provide an accurate observation, ensuring that measurements are always made under the same conditions is less easy. For this reason, surface observations are required to be made at a specified height of the *shade air temperature*, preferably over grass. This specification is designed to ensure that what is measured is the climatologically important parameter – the temperature of the air – and not the capacity of the thermometer to absorb sunlight or the potential of the ground to heat up and so influence what is observed. To achieve this objective thermometers should be mounted in a well-ventilated, louvered, white shelter which prevents either direct sunlight or terrestrial radiation from the ground reaching the instruments. The standard design for such an installation is known as a *Stevenson shelter*.

Even where there are well-maintained temperature records, other requirements can cause problems. Because regular human observations in remote locations are expensive, there are strong economic pressures to use automatic systems which can be interrogated electronically. Although, electrical measuring devices (thermistors) are capable of accurate and reliable measurements, the switch to such instruments can produce significant shifts in observed temperatures (Fig. 6.1). Unless there is careful calibration of the change over, these instrumental effects will go undetected and seriously prejudice the quality of the observations.

Although modern measurements are made under standardised conditions, many earlier observations were made in less uniform arrangements. So it requires careful detective work to construct reliable time series which enable earlier measurements to be compared with current figures. In addition, while the siting of modern instruments can be checked, changes in land use over longer periods can exert important influences on local temperatures. In particular, many early observations were made in the vicinity of towns. As these grew they developed an *urban heat island* which has produced a marked rise in observed temperatures in city centres that could be interpreted by the unwary as climatic warming. Although this warming is real, the fact that urban areas cover so little of the Earth's land surface means the results of urban sites can be misleading. So observations have to be checked to understand what processes have been at work. Similarly, the movement of an observation site from a city centre to the

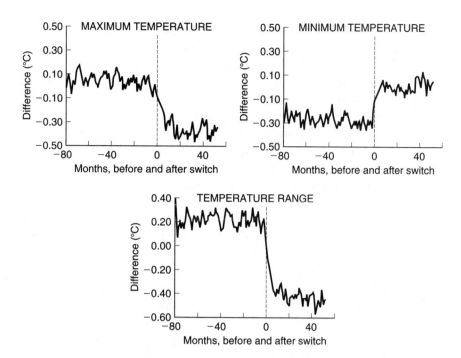

Figure 6.1 Average differences of temperature between stations prior and subsequent to the switch from an alcohol maximum–minimum thermometer in a wooden shelter to a thermistor in a plastic shelter (Karl et al., 1995, Fig. 13).

local airport could lead to a significant shift which must also be carefully calibrated.

The other problem with standard land surface temperature measurements is that, although they have been made in a few places for up to 300 years, they do not provide an adequate coverage of the Earth's surface. Until the late nineteenth century only parts of the northern hemisphere had any semblance of a network of observations. The remaining gaps have been largely filled since, but it requires careful interpolation to build up the record of global land temperatures going back to the middle of the nineteenth century (see Fig. 4.23).

Using temperature measurements at sea have a comparable but different set of challenges. There are a large number of measurements of both air temperatures and surface water temperatures available from the middle of the nineteenth century. Air temperatures made during the daytime are now regarded as unreliable because sunlight absorbed by the decking produced exaggerated figures. But, because the difference between day and night temperatures is much smaller at sea, nighttime observations provide a valuable source of climatic information.

Measurements of sea surface temperatures provide a different set of challenges. In the nineteenth century many ships kept records of the temperature of water collected in buckets over the side of the ship. To use these observations, corrections need to be made for whether the samples were collected in wooden or uninsulated canvas buckets, as the latter cooled more quickly. A bigger puzzle was a sudden jump in temperature of nearly 0.5 °C between 1940 and 1941. It has been established that this change was due to a sudden undocumented shift from using canvas buckets to measuring the temperature at engine intake. This change took place because during wartime using lights to make measurements over the side at night was too dangerous. Applying these corrections to some eighty million observations led to the conclusion that observations before 1941 had to be corrected by an amount that rises from +0.13 °C in 1856, when wooden buckets were widely used, to +0.42 °C in 1940, when canvas buckets were standard. Given that the overall warming of global SSTs since the mid-nineteenth century amounts to about 0.5 °C (Fig. 6.2), these corrections are a major adjustment and show how essential it is to standardise instrumental observations.

The other limitation with SST measurements is the gaps in their geographical coverage. Away from the main shipping lanes, there are many areas with few data, especially in the early part of the record. Coverage was affected by two World Wars and changing trade routes (e.g. the opening of the Suez and Panama canals in 1869 and 1914 respectively). Even in recent decades there is virtually no data for the Southern Ocean south of

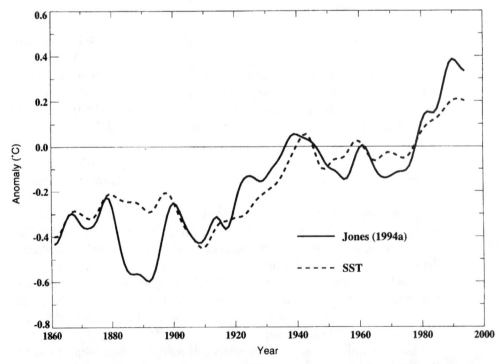

Figure 6.2 Global land-surface air temperature anomalies (°C) 1861–1994, relative to 1961–1990 (solid line) and sea-surface temperature anomalies (broken line) (IPCC, 1995, Fig. 3.3d).

45° S. Given that the oceans cover 71% of the Earth's surface, this is a major deficiency. While additional records may be discovered, there is no prospect of many of these gaps being filled in. So all that can be done is to extrapolate the available data on the basis of known temperature patterns to produce an agreed series for global temperatures.

The global record extending back to 1860 (see Fig. 4.23) combines measurements on both land and sea. The principal uncertainty in the land surface temperature observations is the effect of urbanisation. Overall, out of the observed warming of around 0.5 °C during the twentieth century, less than 0.05 °C can be attributed to urban influences. The incorporation of marine night-time temperatures and SSTs into the global temperature records introduces additional uncertainties. The early data has limitations, but from the beginning of the twentieth century, the estimated errors for the decadal SST anomalies are less than 0.1 °C for those areas with adequate data (about 60% of the world's oceans). The warming observed in these observations at sea is slightly less than detected over the land (see Fig. 6.2). But the overall trend is broadly similar, and when these two sets of measurements are combined they produce the agreed curve for global warming (see Fig. 4.23(a)). This analysis also shows broadly the same trend for both hemispheres (see Fig. 4.23(b) and 4.23(c)), and confirms that the warming is a global phenomenon. The total uncertainty in measurements is reckoned to be ±0.15 °C and this leads to the consensus view of 0.3 to 0.6 °C for the value global near-surface warming since the late nineteenth century.

Using instrumental observations before the middle of the nineteenth century poses additional challenges. Lengthier temperature records are available for Europe and eastern North America. The longest of these is the series that has been produced for rural sites in central England. This series is a classic example of the care that is needed to detect climate change. It was the product of many years of diligent scholarship by the late Professor Gordon Manley and involved a number of interlinked efforts. The first task was to search out and bring together all the records that had been accumulated by a bewilderingly diverse array of amateur observers before the days of official meteorology.

The gathering together of the records was only the start of the analytical problems. First, there was the question of how the measurements were made. Back to the early nineteenth century the combination of reasonable standard observations plus sufficient numbers of overlapping records enabled useful checks to be made of the reliability of the observations and adjustments made for, say, measurements at different times of the day. But earlier records posed greater problems. Before 1760, some of the best-kept records depended on having thermometers exposed in well-ventilated north-facing fireless rooms. A further complication was the change from the Julian to the Gregorian calendar in 1752. Prior to this date it was not possible to compare monthly means with those of England today, neither could they be compared with those of contemporary western Europe unless there were daily observations, as by 1752 the difference in the calendars amounted to 11 days.

What this series shows is that the annual temperature in central England has risen by around 0.7 °C since the end of the eighteenth century (see Fig. 4.20). Most of this warming is concentrated in the winter half of the year (October to March), while there has been little warming in the summer half of the year. Other temperature records for Europe tell roughly the same story. The winter temperature record for Holland can be extended back to 1634 using records of canals freezing and clearly shows a marked warming trend (Fig. 6.3). All sites show a warming since the mid-nineteenth century. The warming is least marked in summer, which was, if anything, warmer in the eighteenth century in eastern Europe than in recent decades.

The other striking feature of these temperature records is the occurrence of consistently cool and warm decades. The 1740s, 1780s, 1800s and 1810s, and the late 1830s and 1840s were consistently cold in Europe. Only the 1820s can be defined as being warm in Europe. In North America the 1780s, 1810s and 1930s were on the cold side, while the 1800s, 1820s and 1840s were inclined to warmth. But, this partial agreement should not be interpreted as being evidence of hemispheric trends. Analysis of twentieth century temperature measurements show that no one region in the Northern Hemisphere is representative of hemispheric-wide conditions. There is no

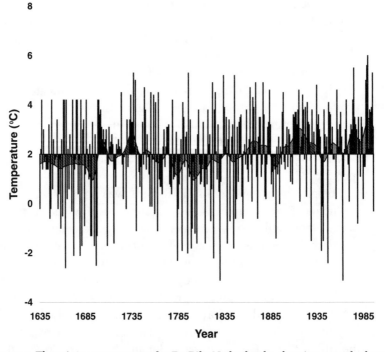

Figure 6.3 The winter temperature for De Bilt, Netherlands, showing a marked warming trend since the late seventeenth century (the record for the period prior to 1705 is based on records of the frequency and duration of the freezing of canals obtained from records of trading activities along the waterways) (Burroughs, 1997, Fig. 5.5).

short cut: the only reliable way to measure hemispheric or global trends is to have adequate coverage of all parts of the world.

A similar set of challenges has to be confronted in analysing other instrumental records. Rainfall measurements on land have been made for at least as long as temperature observations. Here again there are problems with the reliability of early observations, many of which probably understate rainfall amounts because of the siting of gauges. This limitation is the result of any interference of wind eddies near the gauge that is capable of seriously distorting the amount of rain caught. These problems are compounded by growth of vegetation and/or building development around the site. Furthermore, early measurements of snowfall did not catch much of the precipitation. There are also issues with the change of design of instruments. All records in the former Soviet Union must be modified for changes in gauge design in the 1960s.

The other challenge is that rainfall is spatially much more variable than temperature and so in measuring long-term trends requires observations from a large number of sites. Where possible, trends are defined in terms of a number of sites covering a region which has a coherent precipitation pattern (i.e. is affected by broadly the same weather systems). For instance, in the case of the England and Wales series the standard error in the annual figures is more than halved in increasing the number of sites used from five to over twenty (Fig. 6.4). So only where there are copious reliable records of rainfall is it possible to draw conclusions about long-term changes.

Against this background it is hardly surprising that attempts to measure global trends in precipitation amounts are bedevilled by uncertainty. So, far

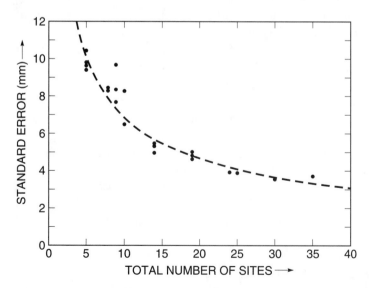

Figure 6.4 The standard error of the estimate of the annual England and Wales precipitation with site density (Wigley, Lough & Jones, 1984; with permission of the Royal Meteorological Society).

less progress has been made in identifying how precipitation has changed with the rising temperatures this century. Prior to the 1970s, measurements are restricted to land areas but this is changing with the advent of satellite measurements (see Section 6.2). Indeed there is no accepted value for the average annual global precipitation. Various estimates since 1960 have varied from 784 to 1041 mm per year with little progress towards agreement.

Over land areas some estimates have been made for changes in precipitation for different parts of the globe. But it is a measure of the scale of the problem that there is no agreement as to whether the continents are getting wetter or drier. There is, however, a consensus on how things have changed latitudinally in the Northern Hemisphere this century. At high latitudes (55–85° N) there had been a rising trend, although part of this may be due to improved instrument design. In mid-latitudes in both hemispheres there is no appreciable trend. The most striking conclusion is the drying of the northern subtropics (10–30° N). This trend is dominated by the desiccation in the Sahel region of Africa, which experienced a 25% decline in successive 30-year periods (1931–60 and 1961–90) (see Fig. 5.7). Elsewhere in the tropics and subtropics there is little evidence of long-term changes. For instance, the trend in the all-India summer monsoon index (see Fig. 3.19), if anything, shows a hint of a wetter trend, and less variability in recent decades compared with the period before 1920.

At the regional level, there are many lengthy precipitation records available. For a few places in western Europe these extend back to the early eighteenth century and for eastern North America from the early nineteenth century onwards. But only limited conclusions can be drawn from these. There is no clear trend to wetter or drier conditions. In England and Wales, where reliable records extend back to the mid-eighteenth century, there is, however, some evidence of a shift towards wetter winters and drier summers. In the United States the 1830s and the 1850s to 1880s tended to be wet and there has been a marked increase in precipitation during the twentieth century (Fig. 6.5). But these fluctuations provide little insight into the processes at work.

As for other forms of instrumental observations, the principal shortcoming is that they have only been maintained extensively for a relatively limited period. So, while local effects can be measured (e.g. changes in sunshine amounts in urban areas provide a good indication of how increases or decreases in pollution levels are affecting the local climate), the absence of comprehensive regional and global networks mean little can be said about how other meteorological parameters have varied in the past, in spite of the existence of a considerable quantity of instrumental data.

6.2   SATELLITE MEASUREMENTS

Since the beginning of the 1960s, weather satellites have offered the potential for global observations of the Earth's weather. Early satellites provided

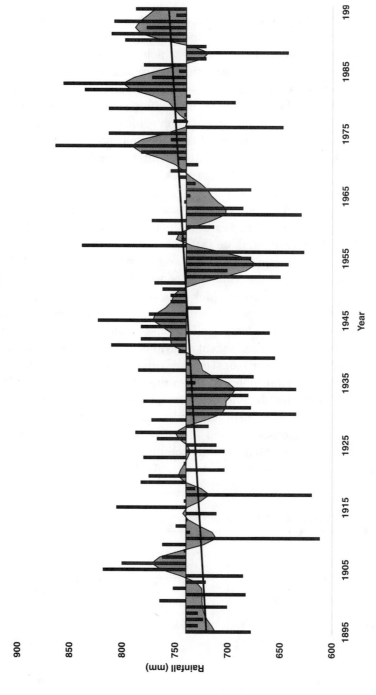

Figure 6.5 Changes in the estimated annual precipitation averaged across the USA since 1895, together with a nine-year binomial smoothing of the data, showing a marked rise over the period (data from NOAA).

147

only rudimentary measurements of cloud cover and temperatures of the surface and cloud tops. This was expanded in the 1970s with the development of infrared radiometers which could analyse the amount of terrestrial radiation upwelling from the surface and the atmosphere at different wavelengths (see Fig. 2.3) to make measurements of temperatures at different levels in the atmosphere. In addition, microwave radiometers were developed which worked on the same principles, and had the huge advantage of seeing through clouds. This meant that not only could they measure temperatures in cloudy conditions, but also make accurate observations of the extent of snow cover and pack-ice in polar regions which are frequently cloud-covered.

Although satellites offer the prospect of global coverage of climate change, when it comes to temperature measurements, they have two basic limitations. First, there is the obvious fact that they have only been making observations for a short time and so can only really contribute to the future monitoring of the climate. Secondly, radiometers measure upwelling radiation from a relatively thick slice of the atmosphere and so their results are not directly comparable with surface results and hence require careful calibration. Nonetheless, satellite measurements of the lower troposphere have already made a significant contribution to the analysis of global warming in recent years (Fig. 6.6). They are capable of making accurate global measurements of atmospheric temperature. But they produce different results for warming trends since 1979 as compared with surface observations. The explanation for the difference lies, in part, in which levels of the atmosphere are being observed, and in the regional nature of the warming since the 1970s. Even after careful examination by the climatological community and a number of corrections, this discrepancy does, however, remain an

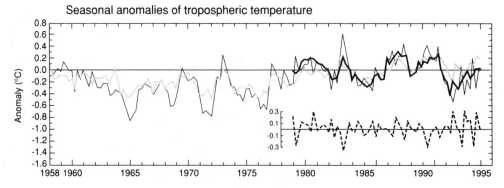

Figure 6.6 Seasonal global temperature anomalies (°C) relative to 1979–94 average, for the 850–300 hPa layer from radiosondes (light solid line) and for the troposphere from satellite/microwave radiometers (heavy solid line). Dashed line (inset) is the difference between observations from satellites and radiosondes. Surface temperatures (shaded line) have been added for comparison with the tropospheric temperatures (IPCC, 1995, Fig. 3.7).

unresolved issue in establishing the scale and causes of current global warming.

The most significant contribution of satellites is in monitoring the change in climatic parameters which were not observed before. Microwave radiometers now regularly measure the extent of snow cover in the northern hemisphere (see Fig. 3.11) and sea ice in polar regions (Fig. 6.7). In the case of snow there are large, short-term fluctuations and some evidence that snow cover in the northern hemisphere has declined in line with global warming. This is hardly surprising given the rise in temperature in the winter half of the year across the northern continents since the 1970s. What is not clear is whether the decline has acted as a positive feedback to amplify the warming or is merely a symptom of the stronger westerly circulation in the late 1980s and early 1990s. During the rest of the 1990s there was a slight rise in the figures, most notably in 1996 and 1997, which may be related to the decline in the strength of westerly circulation in these years (see also analysis of the North Atlantic Oscillation in Section 3.7).

In the case of sea ice, the measurements have transformed the monitoring of polar regions. The figures for the period November 1978 to December 1996 show that the extent of ice in the northern hemisphere has decreased by a rate of 2.9% per decade, while around Antarctica it has increased at a rate of 1.3% per decade. The short duration of these records exposes the frustrations of drawing inferences about longer term fluctuations. Estimates of earlier shifts in the extent of sea ice depend on sporadic observations from occasional scientific expeditions and the interpretation of other data such as the records of the whaling industry. In the Antarctic these records suggest that between the late 1950s and the early 1970s there was a decline of some 25% in the extent of the ice (a reduction in area of 5.65 million km^2), which far exceeds the increase in area that has occurred since around 1976 (see Fig. 6.7(b)). While such observations provide valuable clues about longer term trends, it is hard to be certain that they have captured an accurate picture of what was going on at the time. They do, however, show how essential it is to build up more extensive observations using satellites before jumping to conclusions about the significance of recent trends.

Microwave radiometers can also detect rainfall, so, for the first time, there is a prospect of measuring rainfall over the oceans. Although these observations will require careful calibration before any inferences can be drawn about trends in rainfall, this technology offers the possibility of monitoring precipitation on a global basis. Nevertheless, detailed results of monthly global rainfall measurements are now being published and, over time, they will provide a much better indication of how regional and global rainfall patterns vary in the longer term.

The measurement of cloud-cover is another frustrating area. Since the first weather satellite was launched in 1960, the images obtained have provided large amounts of information about clouds. But, attempts to produce analyses of changing cloud cover have produced equivocal

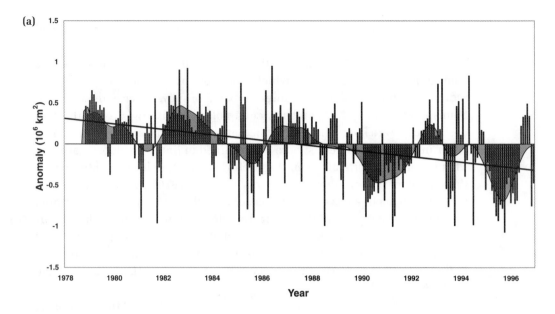

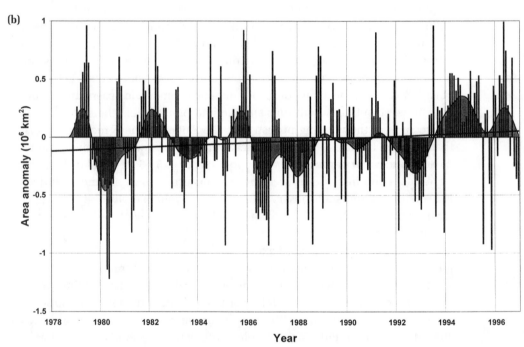

Figure 6.7 Sea ice extent anomalies relative to 1978–96 for **(a)** the northern hemisphere and **(b)** the southern hemisphere. Data from NOAA (USA). Smooth lines generated from a 21-month binomial filter applied to the monthly anomalies to filter out periodicities of twelve months or less.

results. In part, this is a problem of drift. Instruments have become less sensitive to the amount of light reflected from clouds over their lifetime. In addition, any analysis has to discriminate accurately between cloud types and their altitude and many observations are incapable of providing this information. As a result, while results from Russian satellites suggest significant interannual and interdecadal fluctuations in cloud amounts over the last 30 years, observations by US satellites indicate much smaller change. So no clear statement can be made about global cloud cover trends. Surface measurements over land suggest an increase of a few percent during the last 100 years. This increase is closely related to the decrease in diurnal temperature range which has been observed over the same period (see Section 4.10).

6.3   HISTORICAL RECORDS

Where instrumental observations do not exist, information on the climate can sometimes be extracted from historical records, including personal diaries, of both the weather and weather-related activities. But to make a significant contribution to quantitative measurements they must meet sterner tests. Nonetheless, in Europe and China detailed reports of many annual events can provide valuable additional insights. In particular, this is possible where historical records overlap instrumental observations and so can be calibrated against these numbers. But they have limitations, notably that they relate most frequently to agricultural activity. For example, the price of wheat is a good indicator of the abundance of the harvest, as bumper years led to low prices and dearth produced high prices. But, these figures relate only to the summer half of the year and are a complex product of temperature and rainfall. This means that, while the extreme years stand out, many of the fluctuations cannot be attributed to one or other variable. A better measure is the date at which grapes were picked to make wine – the *wine harvest date*. This is a good measure of the temperature of the growing season (April to September).

The best known example of wine harvest dates has been built up from records kept in various parts of northern France and adjacent wine-growing regions by the French historian Emmanuel Le Roy Ladurie and co-workers. These cover the period 1484 to 1879. The extent to which they provide a direct measure of the temperature in the growing season is shown in Fig. 6.8 by comparing annual harvest dates with temperature figures for April to September recorded at De Bilt, Netherlands from 1705 onwards. These show that the correlation between the figures over the period 1705 to 1879 (−0.71) is highly significant (see Box 7.2), and so they do provide a guide to temperature in the summer half of the year. Longer term fluctuations do, however, have to be viewed with caution. The period of later harvest dates in the late eighteenth century and early nineteenth century, which

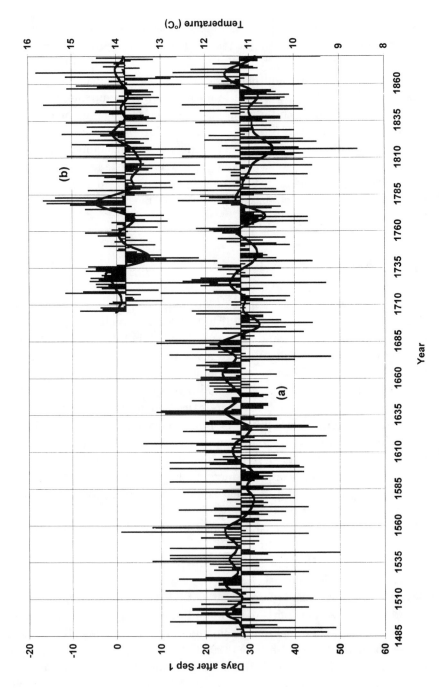

Figure 6.8 The variation in **(a)** the annual average date of the wine harvest in northern France between 1484 and 1879 (see also Fig. 4.19) compared with **(b)** the temperature during the growing season (April to September) measured at De Bilt, Netherlands between 1705 and 1879.

is not evident in the De Bilt temperature record, has more to do with changing fashions in the sweetness of wine: the later the harvest the sweeter the wine.

The limitation of many other historical records is that often they only note exceptional events. For example, as noted in Chapter 4, the incidence of Frost Fairs on the Thames implies colder winters in London, but does not give a measure of how much colder they were in the seventeenth century. With a thorough examination of all the extremes recorded it is sometimes possible, however, to build up a more complete picture of notable changes. A good example of how this approach can produce a lot of information is the work of Christian Pfister on changes of temperature and rainfall in Switzerland since the early sixteenth century (see Section 4.9). But in many cases the records are few and far between and then great care has to be exercised in not attaching too much importance to fluctuations in rare events. Only where there are sufficient observations to construct statistically significant series can these records be used to draw conclusions on climate change. Otherwise, their principal value is to be used in conjunction with proxy data (see next section) to build up a more complete picture of change.

6.4   PROXY MEASUREMENTS

Where there is neither instrumental observations nor historical records, we have to rely on indirect measurements of climate change. For the vast span of the Earth's history these are the only way to find out how the climate has changed. Fortunately, there are a wide variety of techniques for approaching this task. Before examining these in more detail it is illuminating to establish the strengths and weaknesses of using indirect (proxy) data to infer climatic conditions in the past. The products of this analysis are often termed *palaeoclimatic data*.

The first point to emphasise is that most proxy data is rarely ever a direct measure of a single meteorological parameter. For instance, the width of tree rings is a function of both temperature and rainfall over not only the growing season, but also on ground water levels which may relate to rainfall in earlier seasons. Only where the trees are growing near their climatic limit can most of the growth be attributed to a single parameter (e.g. summer temperature). Similarly, ice-core data, apart from the amount of snow accumulating in annual layers being a direct measure of precipitation, other properties are a measure of processes taking place far away from where the snow fell. As for many other records (e.g. analysis of the pollen content and the thickness of annual layers in lake sediments or the foraminifera deposited in ocean sediments), the climatic inferences that can be drawn from changes depend on subtle links between what is being measured and how it was influenced by climate change at the time it was laid down.

The second limitation is that in some sources, information may be blurred by natural processes. For instance the resolution of events in deep ocean sediments is affected by three things: the slow rate of deposition; the fact that this deposition rate may have varied with time; and the possibility that burrowing creatures may have mixed up the layers as they formed. All of these effects mean only changes on timescales longer than around a thousand years can be resolved by these measurements. If the consequence of any churning is to transport microfossils into different layers it can make the interpretation of mass extinctions (see Section 5.3) much more difficult. The transport of fossils downwards into older sediments, which predate some assumed extinction horizon (e.g. the Cretaceous–Tertiary Boundary) is often termed *backwards smearing*, whereas the lifting of other fossils above their horizon is termed *forward smearing*. These processes make it much harder to tell whether relatively rapid changes in population levels were the product of a catastrophic event (see Section 8.10), or took place more gradually.

A third problem is that some records have major interruptions. Dating the organic debris in the terminal moraines of glaciers provides an insight to the timing of their major advances. But, there is no way of knowing whether intermediate, less substantial advances occurred as all evidence of these are scrubbed out by later surges. Similarly, the early over-simplified model of the four major ice ages in the Pleistocene (see Section 4.4) was a result of the fact that the geological evidence of more rapid fluctuations had been erased in those parts of the world where the original work on these events was carried out.

Finally, in many cases, proxy measurements are only available in limited locations, which makes it difficult to draw generalised conclusions about change. Tree-ring or ice-core data can only be found in some particular parts of the world. Many fossil assemblages are the product of exceptional circumstances which may well mean that the populations preserved are not representative of the conditions prevailing more widely at the time. Only in the case of ocean sediments do we have sufficient data to provide something approaching a truly global picture of certain types of change.

6.4.1   Tree Rings

The response of trees to climatic fluctuations is recorded in the annual growth rings they produce. The most direct measure is in terms of the thickness of the rings which not only provides insights into the climate, but also provides the basis of dating ancient trees (see Section 6.5). The challenge facing palaeoclimatologists is to establish what combination of temperature, rainfall and soil-moisture levels led to changes in ring thickness or changes in the wood structure within rings. This is achieved by comparing the observed ring widths in a locality for the period since standard meteorological measurements started in the area. Statistical analysis of the correlation between various meteorological parameters (e.g. summer temperatures or annual rainfall) then establish how much of the variance can be attributed to different climatic factors. This calculation is then used to reconstruct the variation of the given parameter to check how effective the use of the tree-ring data is in representing the past (Fig. 6.9). On this basis, the established statistical link can be applied to earlier tree-ring data to establish how much the climate changed from year to year before instrumental observations were available.

Where trees are near their climatic limits of either rainfall or adequate summer warmth then it is possible to attribute most of the variability in ring width to one or other meteorological parameter. In many parts of the world ring-width fluctuations are, however, due to a combination of factors which make trees grow faster or slower than normal in any year. So it is only possible to draw more general conclusions about how a combination of weather conditions altered tree growth. Moreover, in tropical regions, where there is no pronounced annual cycle in growth, it is not possible to obtain information about climate change.

More detailed information can be obtained from analysis of the structure of the wood within individual rings. Such parameters as maximum latewood density, minimum earlywood density and width of early and latewood growth can provide insight into weather fluctuations within individual seasons. For instance, stunted early growth could indicate late cold springs while lack of late growth could reflect cool, short summers. Also the isotope ratios of the wood (see Box 6.1) should provide additional information about the environmental conditions at the time of growth. In

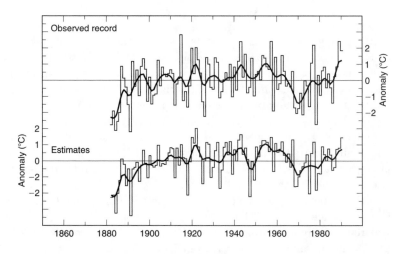

Figure 6.9 Actual and estimated July/August temperatures for northern Fennoscandia, showing how the calibration of tree-ring widths against instrumental observations can be checked by reconstructing the temperature record on the basis of the established relationship. The estimates are based on the regression equation calibrated over the period 1876–1975. The smoothed curve shows ten-year filtered values (with kind permission of K. Briffa, Climatic Research Unit, University of East Anglia).

BOX 6.1   INFORMATION FROM ISOTOPE RATIOS

Many elements exist in a number of stable isotopic forms, each having a different atomic weight. This weight difference between isotopes (e.g. hydrogen [H] and deuterium [D], carbon-12 [^{12}C] and carbon-13 [^{13}C] and oxygen-16 [^{16}O] and oxygen-18 [^{18}O]) alters the physical properties of molecules containing different isotopes. For instance, water vapour molecules containing deuterium($HD^{16}O$) will evaporate a little less readily than normal water ($H_2{}^{16}O$), as will water containing ^{18}O ($H_2{}^{18}O$). Similarly, the rate at which different isotopes of hydrogen are taken up in the formation of wood are likely to be dependent on the temperature. This means that measuring the ratio of certain isotopes in, say, ice cores, ocean sediments and tree rings can provide information on the physical conditions at the time these materials were formed.

In the case of isotopic fractionation in natural water, three principal molecular species ($H_2{}^{16}O$, $HD^{16}O$ and $H_2{}^{18}O$) matter most. In order to describe completely the processes involved, all three species must be considered: so both the difference of deuterium (δD) and of oxygen-18 ($\delta^{18}O$) in any sample from normal are measured. Both may be expressed as deviations per mil (‰) from the Standard Mean Ocean Water (SMOW), where

$$\delta D = 1000 \left\{ \frac{HD^{16}O/H_2{}^{16}O_{\text{sample}}}{HD^{16}O/H_2{}^{16}O_{\text{SMOW}}} \right\} - 1$$

and

$$\delta^{18}O = 1000 \left\{ \frac{H_2{}^{18}O/H_2{}^{16}O_{\text{sample}}}{H_2{}^{18}O/H_2{}^{16}O_{\text{SMOW}}} \right\} - 1.$$

Expressing these deviations in terms of ‰ produces manageable units as $H_2{}^{16}O$ makes up 99.76% of SMOW while $H_2{}^{18}O$ constitutes 0.2% and HDO makes up 0.03%. Most of the water vapour precipitated on the ice sheets of Antarctica and Greenland originates from the oceans in mid-latitudes. The process of evaporation depletes the vapour of about 10‰ of $H_2{}^{18}O$. Then as the vapour is precipitated as rain or snow it is further depleted of $H_2{}^{18}O$ because the isotope condenses out preferentially. Although some additional water vapour will be added from the oceans at higher latitudes, the remaining vapour is increasingly depleted as it rises over the ice sheets, and the value of $\delta^{18}O$ in the snow is a measure of the temperature at which it precipitated. For snow falling on the Greenland ice sheet $\delta^{18}O$ typically ranges from −23 to −38‰ and for Antarctica the range is from around −18 to −60‰. The process for depletion of HDO is the same and the value of δD is given by the expression

$$\delta D = 8\delta^{18}O + 10‰.$$

Because the value of $\delta^{18}O$ or δD in ice cores is related to the route by which the vapour followed, it is not a precise measure of the temperature where the snow fell. So, while changes to these parameters provide a good indication of how temperature has changed over time, they are not an absolute measurement and so must be treated with care. This challenge of identifying which physical factor was responsible for the observed change in isotope ratios assumes greater significance with other measurements. The changes in the $^{16}O/^{18}O$ ratio in the shells of benthonic Foraminifera reflect how the amount of these isotopes locked up in the ice caps fluctuated between glacial and interglacial periods. Because the ice will be depleted in $H_2{}^{18}O$, the levels of this molecular species in the oceans will increase as the ice volume increases. This is a measure of ice volume which can be translated into an indication of global temperatures. But when it comes to measuring the H/D, $^{12}C/^{13}C$, or $^{18}O/^{16}O$ ratios in tree rings, and other organic material, the temperature relationship is more complicated. Although there appear to be identifiable temperature relationships between the observed isotope ratios and ambient temperatures, there is considerable scientific argument as to what precisely is being measured.

practice, in all these areas only relatively little work has been done and, in a number of cases, has produced equivocal results. So, although this type of investigation has considerable potential, more research is needed before any of these techniques can become a larger part of the standard armoury of climatologists. One area that has, however, developed well is that the detection of evidence of frost damage in late spring growth is evidence of an exceptional cold spell (see Section 6.5).

One other aspect of tree-ring studies requires careful thought. This is the fact that all analyses have to take account of the fact that as trees mature, their ring widths naturally decrease. In the case of conifers this change tends to be exponential, but in broad-leaved species (e.g. oaks) it is a more complicated process. Standard practice is to calibrate growth curves for given species and then analyse the annual fluctuations about the normal. This process also involves additional smoothing in the overlapping of series from different tree samples and the combination of these analytical approaches does, however, tend to iron out longer term fluctuations which may include climatic information. The resultant series have a very stable appearance (Fig. 6.10) over the longer term, and it is reckoned by these that studies underestimate the climatic variance by about a factor of two and in particular suppress fluctuations on timescales of fifty years and longer.

6.4.2 Ice Cores

Snow deposited on the ice sheets of Antarctica and Greenland, and in glaciers in mountain ranges around the world contains valuable information about the climate at the time it fell. Where there is no appreciable melting in summer, the accumulation of snow, which is compressed to form ice, contains a continuous record of various aspects of climate variability and

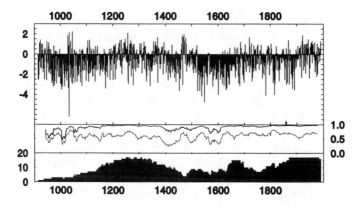

Figure 6.10 The reconstructed July/August temperature for northern Fennoscandia. The values are °C anomalies from the mean for the period 1951–70. The smoothed curve shows ten-year filter values (with kind permission of K. Briffa, Climatic Research Unit, University of East Anglia).

climate change. This includes evidence of changes in temperature, the amount of snow that fell each year, the amount of dust transported from lower latitudes, fall-out from major volcanoes and the composition of air bubbles trapped in the ice. The best results are, however, restricted to Antarctica and Greenland, with more limited results from glaciers and ice caps elsewhere around the world (see Fig. 4.21). Hence, many features of regional climate change are not covered.

The most direct measure of the climate is obtained from the isotopic composition of the water molecules in the ice, and, in particular, the ratio of the stable isotopes of oxygen (^{16}O and ^{18}O) (see Box 6.1). By measuring this ratio it is possible to observe both the annual cycle in the temperature and longer term changes in temperature (Fig. 6.11). Recent ice cores in Greenland have been able to identify annual layers going back some 15,000 years, and measure variations in snowfall over this period, although there are problems in identifying every annual layer and so absolute dating lacks the precision of tree-ring analysis. Dating of older layers, which no longer contain clear evidence of annual variations, depends on making assumptions about the flow of the ice sheet.

In Antarctica where the rate of snow accumulation is much slower, it is not possible to measure annual layers for more than a few hundred years, and most dating depends on the behaviour of the ice sheet. This does not cause major problems for dating back to around 100,000 years ago for the best sites near the centre of the ice sheet in Greenland, while in East Antarctica a record of changes over more than 250,000 years has been built up (Fig. 6.12). But as the core approaches bedrock, the risk of the ice being distorted by the topography of the bedrock in its flow outwards from the centre of the ice sheet can cause major discontinuities in the layers and make the interpretation of change more difficult, if not impossible.

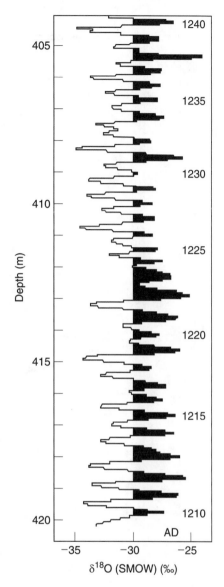

Figure 6.11 An example of the measurement of the $^{18}O/^{16}O$ isotope ratio in an ice core from the Greenland ice sheet, showing the annual layers (source: Wigley, Ingram & Farmer, 1981, Fig. 5.5).

The dust content of the ice provides a measure of atmospheric circulation. Stronger winds in mid-latitudes increase the amount of continental dust stirred up and transported to high latitudes. The changes in amount of dust fluctuate sharply with shifts in temperature, showing that the variations in isotope ratios are also reflected in switches in atmospheric circulation patterns.

Measurements of the acidity of different layers of the ice core provides clear evidence of major volcanoes. Because the most climatically important volcanoes are those which inject large quantities of sulphur compounds into the stratosphere to form long-lasting sulphuric acid aerosols (see Section 8.4), the acidity of precipitation at high latitudes is a good measure of the climatic impact of these eruptions (Fig. 6.13).

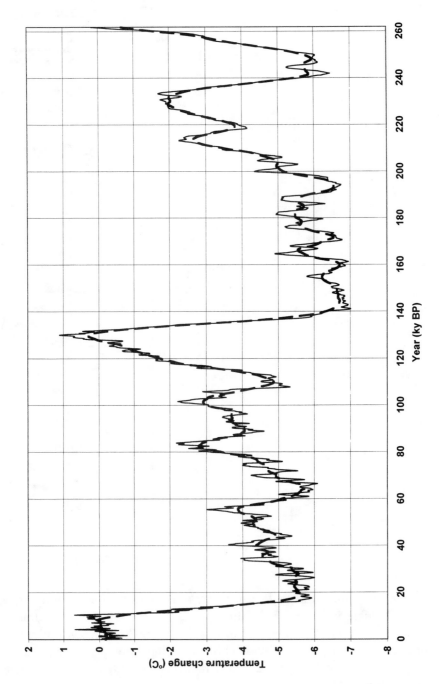

Figure 6.12 The change in temperature derived from the measurement of the hydrogen isotope ratio (H/D) in an ice-core drilled at Vostok in Antarctica, showing how the ice core record extends back 260,000 years covering the last two glacial and interglacial periods (data from NOAA WDC-A).

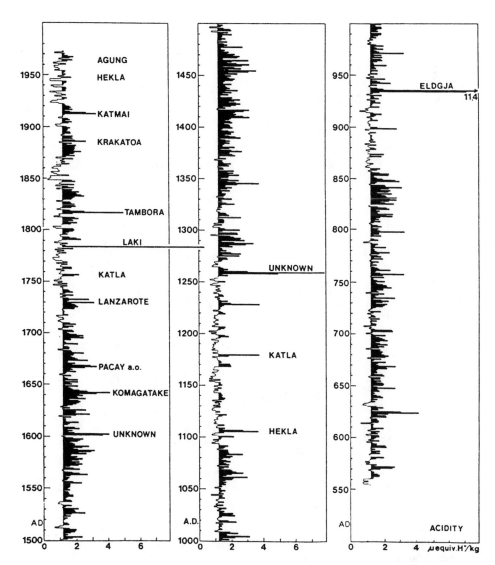

Figure 6.13 Measurements of the acidity of a Greenland ice core showing the peaks that occur after a major volcanic eruption (Hammer, Clausen & Dansgaard, 1980; with permission of Macmillan Magazines Ltd).

The analysis of the trace constituents of the atmosphere in the air bubbles trapped in the ice provides a direct sample of the amount of each gas present in the atmosphere when the snow fell. This gives a history of the past fluctuations of trace gases such as carbon dioxide (CO_2) and methane (CH_4) (Fig. 6.14), and is an indication of how these changes may have contributed to past climate change (see Sections 2.1.3 and 8.9).

6.4.3 Ocean Sediments

The sediments that collect on the bed of the deep oceans provide one of the most valuable sources of information on longer term climate change.

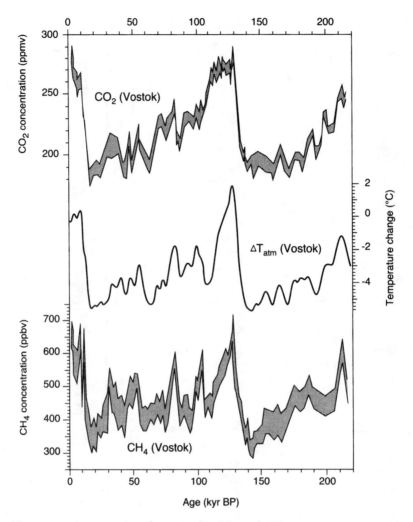

Figure 6.14 A comparison between the CO_2 and CH_4 concentrations in an ice core obtained at Vostok, Antarctica, and the estimated changes in temperature over the last 220,000 years (IPCC, 1994, Fig. 1.6).

Because this ooze is formed largely of the bodies of the fossil shells of both pelagic and benthonic foraminiferal species, it provides a measure of the conditions when these tiny creatures were living in either the surface or deep waters of the oceans. So taking a core down through this sediment which has collected over hundreds of thousands of years provides a record of the species living above a spot in the past. The fact that these sediments occur across the major ocean basins means that it is possible to build up a picture of global changes drawing on a substantial ocean drilling programme over the last three decades.

Two principal factors are used in the analysis of cores. First, the types of species living in the surface at any given time and their relative abundance is a guide to surface temperatures – whether the various species only survive in warm or cold water is a direct measure of the conditions when they were alive. So by mapping the populations of different creatures at different times using cores taken from around the world, it is possible to construct a picture of how the temperature of the oceans' surface waters varied over time. The second factor is the ratio of oxygen isotopes in the calcium carbonate in the skeletons of the foraminifera living in the deep water. Changes in this ratio are a longer term consequence of the fractionation process described in Box 6.1. As the ice sheets in the northern hemisphere grew, the amount of ^{16}O locked in the ice was proportionately greater than the amount of ^{18}O. So the ratio of these oxygen isotopes in the oceans changed as the amount of ice rose and fell, and these variations were reflected in the shells of the foraminifera which formed using oxygen in the oceans, whether they were living close to the surface or deep down (Fig. 6.15). This means that changes in the $^{16}O/^{18}O$ ratio is a measure of the size of the ice sheets at high latitudes, and hence indirectly of fluctuations in global temperatures.

The dating of these sediments is a more complicated issue. Carbon-dating can handle the last 40,000 years or so, and in the longer term magnetic reversals can act as markers. But for the climatically interesting period covering recent ice ages, dating depends initially on assumptions about sedimentation rates. Climatologists are, however, fortunate in this context as the data from a large number of cores shows beyond peradventure that changes on timescales from 10,000 to 100,000 years are dominated by the Earth's orbital parameters (see Section 8.7). This means that while the assumption of constant sedimentation rates is a reasonable first assumption for many cores, once the orbital link has been established, it is possible to refine the timescale to fit the orbital parameters more closely. This 'tune-up' process has led to an agreed dating for the major changes in ocean isotopic ratios, which are a measure of the major stages in the fluctuations in ice cap volumes over the last 500,000 years or so. This is known as the SPECMAP chronology.

6.4.4 Pollen Records

Where sediments have formed in shallow lakes or bogs, the presence of pollen grains can provide similar information on land, as foraminifera provide in the oceans. If these sediments have been laid down in a regular manner, the abundance of pollen from different species of trees and shrubs provide details of climate change. Because different species have distinct climatic ranges, it is possible to interpret their relative abundance in terms of shifts in the local climate. Given that this process can occur almost anywhere on the continents, it means that pollen records have the potential to fill many of the geographical gaps which ice cores and ocean sediment records cannot cover.

Pollen records are of particular historical importance in studying climatic change because of the early work charting the emergence of northerly latitudes from the last glaciation (see Section 4.5). Most pollens (from flowering trees and plants) and spores (principally ferns and mosses) are tiny. Few

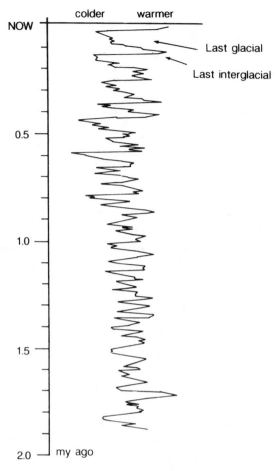

Figure 6.15 An example of the information that can be extracted from ocean sediment records which can be used to draw conclusions about the amount of freshwater locked up in the ice caps. The striking features emerging from a large number of these records are the 'sawtooth' nature of these curves, which reflects the slow build up and rapid collapse of the major ice sheets every 100,000 years or so during the last 800,000 years, and the more rapid fluctuations which occurred before this time (source: Van Andel, 1994, Fig. 4.2).

exceed 100 μm (0.1 mm) in diameter and the majority are around 30 μm. It is the outer portion of the cell (*exine*) which is preserved by means of a waxy coat of material *sporopollenin*. The size and shape of this outer wall, along with the number and distribution of apertures in it are specific to different species and can be readily identified. A typical pollen diagram (Fig. 6.16) will plot the proportion of the principal species found in a stratigraphic sequence and can be used to draw inference about how the climate has changed over the period covered by the record.

In spite of their potential, most climate change studies using pollen records have focused on alterations in regional vegetation since the end of the last ice age. This work has established that these shifts are con-

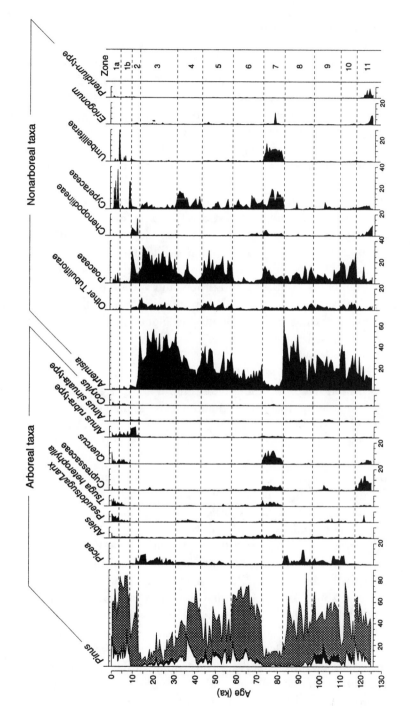

Figure 6.16 An example of the information that can be extracted from pollen stratigraphy, using a core taken from the bed of Carp Lake in the Cascade Range of north-western USA. Note how the changes observed here mirror changes seen in ice core and ocean sediment data (see Fig. 4.13). (Whitlock & Bartlein, 1997; with permission of Macmillan Magazines Ltd.)

trolled by broad global patterns of climate change. For longer term changes there was until recently greater uncertainty as to whether other factors involving the migration and competition of species may have played a more significant role. There are two principal explanations for this uncertainty. First, almost all the longer pollen records related to northern Europe and hence there were doubts about whether the results were representative of global changes. Secondly, many of the cores contained high frequency variations during the last glaciation. These raised doubts about the dating of the strata in the cores and were not readily explained until more recent ice core and ocean sediment data became available (see Fig. 4.13). Recent pollen results obtained from cores drilled in Carp Lake in the Cascade Range in north-west USA (see Fig. 6.16) have confirmed that pollen records can be accurately dated back to the last interglacial, some 125,000 years ago, and provide detailed information about how the climate has changed. Furthermore, the close correlation between these results and those from cores from lakes in Europe, and variations observed in ice cores and ocean sediments means that future research into pollen records will play an increasing part in identifying global patterns of climate change during the last ice age.

6.4.5 Boreholes

Holes drilled deep into the ground offer a completely different and potentially valuable approach to estimating past climates. By making measurements down boreholes, it is possible to relate profiles of rock temperatures with depth to the history of temperature change at the surface. With certain assumptions this change can be converted to the changes in air temperature. In principle, by making measurements to depths of a kilometre or more, this can provide insights into local temperature variation over time intervals of a few hundred years to over a millennium.

Because boreholes are drilled in so many parts of the world, temperature-depth profiles from some 1,000 boreholes have been used for climate research. The geographic coverage is most dense in North America and Europe, but substantial data have also been obtained from Australia, Asia, Africa and South America, and new borehole measurements are continually being obtained.

The calculation of temperature records shows substantial sensitivity to the assumptions made to convert the core temperature profiles to atmospheric temperature changes. The usual initial assumption employed, when setting up the mathematical procedures to convert borehole temperature profiles to temperature changes at the surface, is that surface temperature has remained constant. Then, using certain simplifying assumptions about the geothermal properties of the earth in the area of the core to calculate what the temperature profile should be under isothermal conditions, the deviations from this expected profile can be translated into past changes in the surface temperature.

A number of additional factors do, however, complicate the interpretation of these measurements. First, the temporal resolution decreases as one
goes back in time. Secondly, long-term variations in winter snow cover and
in soil moisture change the sub-surface thermal properties of the ground. In
addition, non-climatic factors such as land-usage changes and natural land
cover variations can affect the response to surface temperature. For instance,
land usage changes may explain why some borehole estimates of warming
in North America are 1–2 °C greater than instrumental records of local air
temperatures show to be the case.

6.4.6 Other Proxy Measurements

Alongside the principal proxy measurements described above, there are a
wide range of others which are used where the circumstances permit.
Wherever a natural process lays down long-term deposits of material at an
approximately constant rate, there is the prospect that some climatic information will be recorded. For instance, the deposition of calcium carbonate
encrustations by running water in caves (*speleothems*) record changes in the
isotopic ratios in precipitation. So, where such encrustations build up over a
long time, they have the potential to provide useful climatic information.
One of the best known examples of this process is a 36-cm layer that built up
in Devils Hole, Nevada, USA, which has provided valuable data for the
period from 500,000 years ago to 60,000 years ago, after which the encrusta-
tion ceased.

In parts of the tropics, where there are sufficient differences in the temperature and sunshine levels at various times of the year certain forms of
coral (e.g. the star coral *Pavona clavus*) produce seasonal growth rings. The
coral grows faster in warm sunny conditions than when it is cooler and more
overcast. The resulting bands of alternating density can be measured using
X-ray techniques and used to draw conclusions about changes in seasonal
water temperature. In addition, the changes in the $^{16}O/^{18}O$ ratio in the coral
is another measure of the temperature at the time of formation. Using these
techniques, measurements of temperature fluctuations in the Galápagos
Islands have been obtained back to AD 1600, which provide an independent
record of ENSO events since then.

Corals can also record variations of freshwater run-off because they
may extract and co-deposit river-borne organic compounds in their
annual growth rings. The amount of organic material from the freshwater
can be measured by recording the fluorescence in the yellow–green part
of the visible spectrum when the coral is illuminated with ultraviolet
(UV) light. Studies inside the Great Barrier Reef off the coast of
Queensland, Australia, have shown that the large slow-growing (5–25 mm
per year) corals of the genus *Porites*, which may live for up to 1,000 years
and grow to 10 m across, provide an accurate record of the amount of run-off
from local rivers. This is a measure of rainfall in tropical Australia, which
provides further evidence of ENSO events. This technique has also been

used to study ancient coral reefs to study the climate associated with past high sea levels.

6.5   DATING

It should be obvious by now that dating proxy records is no easy task. Only where it is possible to build up on unbroken sequence of measurements to the present is it possible to create an absolute chronology. For the rest, there are two principal limitations. First, where there are gaps in the record (e.g. lack of overlapping tree-ring series), some way has to be found to link the measurements with a reliable timescale. Secondly, where there is some physical means of measuring the age of the sample (e.g. carbon dating: see Box 6.3) there are questions of both the accuracy of the technique and the problems of contamination.

The best examples of creating an absolute timescale are achieved with tree ring and ice cores. In the case of ice cores, this is a matter of counting the layers in any given core, but with tree rings it is possible to build up chronologies using different samples that have grown in a given location. Because the annual growth of trees is directly related to fluctuations in the weather, and these are never the same over a long run of years, the variation of tree-ring widths is a unique record of how any tree has responded to local climatic fluctuations in its lifetime. Moreover, because anomalous weather patterns extend over considerable areas, it is possible to use trees from up to several hundred kilometres apart to build up a chronology. So by sampling living trees in a region a series can be extended back several hundred years, or with certain trees much longer (e.g. the Bristlecone pine – *Pinus longaeva* in the White Mountains of California, which can live several thousand years; Fig. 6.17). Records can then be extended back using timber from trees which grew in the locality and were felled long ago and either preserved in buildings and artefacts, or fossilised. By overlapping these samples both with each other, and in the case of relatively recently felled timber with living samples, it has been possible to produce chronologies extending back between eight and nine thousand years in the case of European oaks, and eight thousand years in the case of Bristlecone pines. What is more, by using a large number of trees, it is possible to overcome problems of missing very narrow rings in single samples that experienced particular stress in some years.

Where there are real doubts about either the rate at which sediments or other material has been laid down then the problems of dating become much greater. These challenges were discussed in terms of ocean sediments earlier (see Section 6.4.3). One valuable tool in addressing this issue is to measure the residual magnetic field in the sediments. Because the magnetic polarity of the Earth's magnetic field reverses from time to time, this can leave an indelible record in rocks laid down at the time. On the scale of geological

Figure 6.17 Bristlecone pine (*Pinus longaeva*), native to mountainous areas in the south-western USA. These trees grow at altitudes above about 3,000 m. Because of their extreme longevity and the fact that they live near the limit of their climatic tolerance, they are uniquely important in dendroclimatological studies (photograph Copyright © BBC, London – used in Burroughs 1994, Fig. 4.4).

time, reversals are common, occurring irregularly, but on average about every 700,000 years. Indeed, the last reversal occurred 700,000 years ago, and this marker, together with earlier reversals during the last five million years, which have been dated with considerable accuracy, are central to the accurate dating of ocean sediments.

6.6 ISOTOPIC AGE DATING

Dating rocks using their radioactive content is a simple concept. Certain widely distributed elements (e.g. carbon, potassium, thorium and uranium) occur as radioactive isotopes as well as stable isotopes. Radioactive isotopes decay. (Uranium isotopes decay to lead, potassium-40 (^{40}K) to argon-40 (^{40}A)

and carbon-14 (^{14}C) decays to nitrogen-14 (^{14}N).) The rate of decay is characteristic for each isotope, which cannot be changed by any known force, and is defined as the *half life* – the time taken for half the number of atoms originally present to decay to daughter atoms. So providing we can assume that only the parent radioactive isotope was present when the sediments were laid down, and neither the parent nor the daughter elements have been lost or gained since the mineral was formed, it is possible to calculate its age (see Box 6.2). If the daughter element leaks away the date will be too young; if the parent leaks away the date will be too young.

Carbon-14 (^{14}C) is different. It has not been present since the Earth formed. Indeed it is created continuously in the upper atmosphere by colli-

BOX 6.2 RADIOACTIVE ISOTOPE DATING

In principle, the age of any sample containing a radioactive isotope can be calculated on the assumption that at some point in time the amount of the isotope was fixed. The most direct approach is to estimate the amount of the isotope present now (Q) as a proportion of the amount originally present (Q_0), as the relationship between the two quantities is given by the expression:

$$Q = Q_0 \exp(-t/\tau)$$

where Q is the quantity present at time t, Q_0 is the quantity present at time $t = 0$, and τ is the average life expectancy of the isotope and is the time taken for the amount of the isotope to fall to $1/e$ of its original quantity where e is the natural logarithm (2.718).

The lifetime of the isotope (τ) is related to the half life (t_1) by the expression

$$t_1 = 0 \cdot 693\tau$$

In the case of ^{14}C the value of t_1 is 5,730 years and so τ is 8,267 years and so the age of a sample is given by the expression

$$t = -8267 \ln(Q/Q_0)$$

where ln is the natural logarithm.

So, providing we know what Q_0 was when the sample was formed, measurement of Q will give a direct measurement of its age. In the case of radiocarbon, dating this analysis is limited to values of t less than about 40,000 years, because beyond this the amount of ^{14}C even in a large sample becomes very difficult to measure. Other isotopic combinations can, however, be used to date much older samples. For example, the use of the decay of potassium-40 to argon-40, which is often used to date the solidification of molten lava following volcanic activity, has a half-life of 1,250 million years. This much slower decay process can be used to date samples extending back over much greater lengths of time. But, because of the sensitivity of mass spectrometry to detecting ^{40}A, it is also possible to date eruptions as recent as some 30,000 years ago.

sions between cosmic rays and the nuclei of nitrogen. The ^{14}C created diffuses through the atmosphere as carbon dioxide (CO_2), is dissolved in the oceans, converted by plants into organic matter and ingested by animals. It is capable of decaying to ^{14}N as soon as it is formed, but in living organisms an equilibrium is maintained with the levels in the atmosphere or the oceans. Once the organisms form lasting tissue or when they die, however, the radioactive decay of ^{14}C continues without replenishment. So, providing the amount of ^{14}C in oceans and atmosphere has remained constant it is possible to use the residual radioactivity in samples of dead tissue, wood or shell to determine their age (see Box 6.3).

BOX 6.3 ACCURACY OF CARBON DATING

Because the half life of ^{14}C is 5,730 years (Box 6.2), the carbon-14 atoms in any sample of material will decay at an immutable rate of 1% every eighty-three years. So, at the most basic level, the dating of a sample depends on measuring either the rate at which ^{14}C atoms are decaying, or the proportion of ^{14}C atoms remaining in the sample. If the rate of decay is measured then to obtain an accuracy of $\pm 0.5\%$ in the number of atoms present would require the observation of 40,000 decaying carbon atoms. This would, in principle, allow the sample to be dated to an accuracy of ± 40 years and would require a sample of several grams to be measured for about a day. If, however, the proportion of carbon-14 to carbon-12 atoms in a sample is measured directly in a mass spectrometer, the same accuracy can be achieved in a few hours using a sample of a few milligrams.

In practice, the absolute measurement of the proportion of ^{14}C present in the sample is only one of a number of sources of error in carbon dating. Most significant among these is the fact that the amount of ^{14}C created in the atmosphere has varied over the last 10,000 years as a consequence of changes in the Sun's magnetic field (see Section 8.5). These fluctuations in ^{14}C production have been estimated on the basis of measurements of tree rings. These measurements have produced a calibration curve which compare the actual area of the tree rings with that inferred by measuring the proportion of ^{14}C remaining in a ring of a given age (Fig. B6.1). Two sets of results are shown. The first has been produced by the University of California, at La Jolla, using data from bristlecone pines in the White Mountains of California, and the second by the University of Belfast using European oaks.

These curves show two important effects. First, over much of the calibration period the production of ^{14}C was higher than in the recent past and so the carbon-dating process would, in the absence of the correction obtained from the calibration curve, underestimate the age of the sample. Secondly, the production of ^{14}C clearly fluctuated on timescales of a century or more. In particular, a fluctuation with a periodicity of around 200 years stands out clearly. These fluctuations (often termed *de Vries wiggles*

(continued)

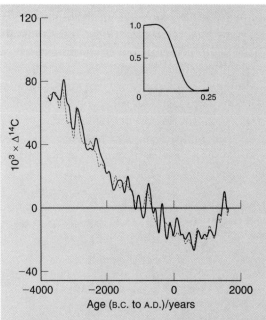

Fig. B6.1 The time sequences of the changes in the atmospheric inventory of ^{14}C, filtered to remove periods less than around 100 years, obtained by the University of California (solid curve) and the University of Belfast (broken line), showing the amount of radiocarbon in the atmosphere declined from around 6000 BC to a broad minimum near AD 500 before rising a little in more recent centuries (Sonnet and Finney, Fig. 1, in Pecker and Runcorn, 1990).

after the scientist who first studied the variations in the calibration curve) mean there are periods when declines in ^{14}C production effectively equal the radioactive decay of ^{14}C. For instance, if, as a result of changing solar activity, the production of ^{14}C declined by 1% over eighty years, this change would be indistinguishable from the same decline in a sample which had died at the beginning of this period. So, there are periods where declining ^{14}C production make it much more difficult, if not impossible, to date samples to an accuracy of better than two or three hundred years in three or four millennia.

Other sources of error are more difficult to identify. Because carbon dating depends on the sample dying or being cut down at some given time we assume we are looking only at the remnants of the ^{14}C that was in the sample at this instant. Any addition of carbon from other sources (e.g. other organic material and, in particular, any recent material which could be affected by such human activities as the combustion of fossil fuels and nuclear testing) will introduce errors. More subtle is where plants have been growing in hard water with particularly high concentrations of calcium carbonate which is depleted in ^{14}C.

The importance of carbon dating for climate studies is in examining organic samples, including some tree-ring series and peat deposits, which are not part of a continuous record. These can be dated directly using radiocarbon measurements. The accuracy of dating depends on both the technology of both measuring and calibrating carbon-dating systems (Box 6.3), and ensuring that the samples are not contaminated by extraneous sources of carbon of a different age.

One other form of isotope dating, which is growing in climatic importance, uses the ratio of ^{230}Th to ^{234}U in the series of decay products from uranium to lead. This method depends on finding materials which form a *closed system* in which the amount of uranium incorporated at some given time is known, and there is no external source of thorium. Living coral absorbs about three parts per million of uranium from sea water, and this proportion has not changed appreciably over the timescale for which corals are used in climatic studies. There is negligible incorporation of thorium into the coral. So the ratio of ^{230}Th/^{234}U is a measure of when the coral was formed. This calculation is, however, complicated by ^{234}U having a half life of 245,000 years and ^{230}Th having one of 75,400 years, but this is manageable providing there was no thorium present when the coral was formed. In these circumstances, the thorium will eventually reach an equilibrium level with the uranium. The rate at which the concentration of the daughter product (^{230}Th) approaches equilibrium with the parent (^{234}U) is defined by the half life of the thorium (75,400 years). This technique is particularly well-suited to dating coral stands left by high sea levels during interglacials (see Section 4.4) in the last 250,000 years or so.

Finally, there is one other climatically interesting isotope dating method, which relies on the ratio of the two stable isotopes ^{86}Sr and ^{87}Sr of the element strontium. The isotopes derive from the weathering of continental and oceanic rock and are found everywhere in ocean water in minute quantities. Solutions leached out of continental rocks have a high ^{87}Sr/^{86}Sr ratio, whereas oceanic islands and volcanoes have a low one. The contributions from the two sources have altered over time and the ^{87}Sr/^{86}Sr ratio of ocean water has varied with them. So it has been possible to measure, using rocks dated by other means, how this ratio has changed over time. This agreed timescale can then be used to date calcareous fossils because strontium is chemically similar to calcium and is stored in these shells. The timescale based on the ^{87}Sr/^{86}Sr ratio is used to date fossils laid down over the last 65 Myr.

6.7   SUMMARY

Only by exploiting a wide range of accurate measurements is it possible to build up a coherent picture of climate change. Even so there are substantial gaps in all aspects of our knowledge, and the further we go back in time the greater these become. New technologies (e.g. satellite radiometers) can improve the monitoring of current conditions or the precision of exploiting sources of palaeodata (e.g. accelerator mass spectrometers for improved carbon dating). New sources of past climatic information will be found to fill some of the gaps. This will involve extending the available proxy records of tree rings, ice cores, pollen and ocean sediments and by developing new sources of data (e.g. speleothems, corals and boreholes). Progress in understanding the causes of, and predicting the future course of climate change depends, in part, on them squeezing the most information out of this data. This takes us into the demanding world of statistics.

QUESTIONS

1   Identify the most significant causes of the urban heat island. What checks must be conducted to ensure that temperature readings taken in the vicinity of cities are not distorted by urban development? Are the same checks needed for rainfall measurements and, if so, what corrections have to be applied?

2   Compare and contrast the strengths and weaknesses of various forms of proxy measurement of past climates. In the light of this analysis, identify the most important improvements needed to enable the different forms of measurement to be used together to provide a better picture of climate change.

FURTHER READING

A complete reference list is available at the end of the book but the following is a selection of the best books or articles to follow up particular topics within this chapter. Full details of each reference are to be found in the Bibliography.

Baillie (1995). A fluent and highly readable account of how tree-ring data is analysed and the contribution it can make to various historical studies including climate change.

Fritts (1976). A standard text by a leading authority on the extraction of climatic information from tree rings which provides a comprehensive and informative review of dendroclimatology.

Karl et al. (1995). An illuminating discussion of the many problems that confront climatologists is making effective use of available data and ensuring that various forms of bias and inaccuracy do not creep into the analysis.

Pecker & Runcorn (1990). A set of papers which explore the links between the Earth's climate and variability of the Sun, and provide useful insight into the implications for carbon dating of changes in solar activity.

Robin (1983). A set of papers which provide an excellent introduction to how climatic information can be extracted from ice cores.

Wigley et al. (1981). A series of papers by a number of leading climatologists which provide many useful insights into the discipline that must be exercised in extracting climatic information from historical and proxy data.

CHAPTER SEVEN

STATISTICS, SIGNIFICANCE AND CYCLES

All is flux, nothing stays still

Heraclitus c.540–c.480 BC

Squeezing useful information out of the available data is the essence of unravelling the causes of climate variability and climate change. The challenge is to exploit statistical techniques and tease out significant cycles, shifts or trends in the climate from the sea of noise which washes over all aspects of this subject. This requires a disciplined approach to avoid falling into the trap of attributing too much to what is nothing more than noise because, the one thing that is certain, is that every aspect of the climate fluctuates on every timescale. So what really matters is to define specific meaning to the terms introduced in Chapter 1 (Figs. 1.1 and 1.2) in order to substantiate any conclusions reached about past changes and to provide a benchmark for making predictions about the future.

This process of analysis is not just a matter of identifying significant changes in the climate (e.g. trends, cycles or sudden shifts) but also involves interpreting the properties of the associated variability. The latter can occur on every timescale within the period covered by the observations. The only way to conduct this analysis is to consider the statistical techniques designed to get the best out of noisy data. This involves a wide range of complicated mathematical techniques. These will not be discussed here. Instead we will concentrate on three practical matters. First, there is the problem of dealing with extreme events which are a normal part of climatic variability. Deciding whether the changing incidence of these rare events is a sign of a lasting shift in the climate is a fundamental issue. This leads into the second issue of identifying significant trends in the data as quickly as possible.

This is particularly relevant to interpreting current events and deciding whether the observed behaviours are simply an expression of the natural variability of the climate or evidence of a more significant change. The third area is the techniques used to analyse data to identify the presence of cycles.

7.1   TIME SERIES, SAMPLING AND HARMONIC ANALYSIS

The starting point is the fundamental properties of the data under examination. Many of the examples already considered in earlier chapters provide an indication of the type of data we want. Ideally, it should consist of an accurate measure of whatever parameter (e.g. precipitation, pressure, temperature, wind speed etc.) at equal intervals of time, as often as needed for as long as was required. By 'accurate' we mean that errors in the measurement technique are small compared to the changes in the parameter we wish to study. Such a *time series* would then allow us to draw detailed conclusions about the behaviour of the parameter over the period of the observation (see Box 7.1). But, as we have seen, for all sorts of reasons the record is limited by many practical considerations. On the other hand, where the record is complete, there may be far more data than is needed to consider longer term fluctuations in the climate. In these cases there are good reasons for working with the minimum amount of data required to meet certain criteria to obtain unambiguous information about real changes in the climate.

The simplest of these criteria relate to the sampling interval and the length of the series. Because of the dominant nature of the annual cycle, many series consider annual average figures when considering longer term changes. The same applies to analysis of seasonal trends. Figures for, say, winter temperatures (i.e. the average of daily figures for December, January and February) will be quoted on an annual basis, independently of spring, summer and autumn values. Analysis of such series can tell us nothing about periodic variations shorter than two years in length. At the other extreme, if the length of the record is, say, 200 years, it is impossible to say anything about periodicities longer than this timescale, and difficult to draw reliable conclusions about any periodicity longer than 50 to 100 years.

These limiting conditions arise out of a fundamental mathematical property of time series. The French mathematician, Jean-Baptiste Joseph Fourier demonstrated that any time series consisting of $2N$ equally spaced points can be expressed as the sum of N harmonics of differing amplitudes (see Fig. 7.1). So in the case of a set of 200 successive annual observations this means the series could be expressed as the sum of 100 harmonics, with the first harmonic having a period of 200 years, the second 100 years, the third 66.7 years, and so on to the hundredth harmonic with a period of two years exactly. The process of calculating the amplitude of these harmonics is

BOX 7.1 METEOROLOGICAL TIME SERIES AND VARIANCE

Any regular measurement of a meteorological variable (e.g. pressure, rainfall or temperature) can be expressed as a time series. This series can be defined as:

$$X(t) = X_0, X_1, X_2, \ldots \ldots X_N$$

where X_0, X_1, X_2 etc., are successive observations of the given meteorological parameter at equally spaced intervals at times O, Δt, $2\Delta t$, etc. The entire series consists of $N + 1$ observations and covers a period P ($P = N\Delta t$). Given we are usually concerned with how $X(t)$ varies from the normal, it is standard practice to define the series in terms of variations about the mean value $\bar{X}$, where

$$\bar{X} = \frac{\sum_{n=0}^{n=N} X_n}{(N+1)}$$

So the new series $x(t)$ can be defined as

$$x(t) = (X_0 - \bar{X}), (X_1 - \bar{X}), (X_2 - \bar{X}), \ldots \ldots (X_N - \bar{X})$$
$$x(t) = x_0, x_1, x_2 \ldots \ldots x_n$$

where $x_0, x_1, x_2 \ldots \ldots x_N$ are the deviations of each successive observations about the mean $\bar{X}$, and can be either positive or negative.

The variance of any time series is defined as

$$\sigma^2 = \frac{\sum_{n=0}^{n=N} x_n^2}{(N+1)}$$

This definition of the variance includes all fluctuations about the mean $\bar{X}$ which could cover not only random variations, but also regular changes (e.g. the annual cycle) and long-term trends. Depending on the nature of the statistical studies being conducted, it may be more illuminating to remove these identifiable sources of change from the series, before considering the nature and scale of the residual variance. This process frequently involves the removal of the annual cycle and, in the case of spectral analysis (see Fig. 7.2), may also include removing any identifiable linear trend, especially if there is some doubt about its climatic reality (Box 7.2).

known as *Fourier analysis* and the resulting set of harmonics is known as the *Fourier transform* of the time series. Because the amount of variance in any time series, which is the sum of the squares of the difference of each point in the series from the average value (see Box 7.1), is the standard measure variability, it is usual to consider the square of the amplitude of each harmonic, and the values of these form the *power spectrum* of the variance in the time series.

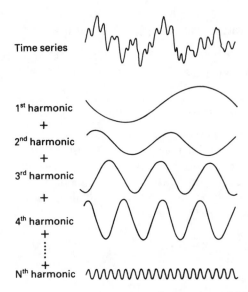

Time series

1st harmonic
+
2nd harmonic
+
3rd harmonic
+
4th harmonic
+
⋮
+
Nth harmonic

Figure 7.1 A time series may be represented by the combination of a set of sine waves (harmonics) of differing amplitudes and phases (Burroughs, 1994, Fig. 2.1).

Computer programs, which can calculate the Fourier transform of lengthy time series, are readily available for many personal computers (PCs). So it is easy enough to explore the harmonic components of time series, and the important discipline is to interpret accurately the meaning of the computed spectra. The easiest way to understand this discipline is to consider some examples of time series and their complementary power spectra. Starting with the most trivial example, the monthly temperature record for a mid-latitude site in the northern hemisphere would be dominated by the annual cycle (Fig. 7.2(a)), where virtually all the variance is in this cycle and the residual scattering of the other lesser components, reflect all the other fluctuations from month to month and year to year.

A slightly less trivial example of a Fourier transform can be found in sunspot numbers. Given that so much of the search for cycles in the weather has been associated with finding links with solar activity, it is a good example to consider. As Section 8.5 describes, sunspot numbers show pronounced cyclic behaviour with the major fluctuations having an approximately eleven-year period. In addition, successive eleven-year peaks show a periodic variation in intensity which reflects a periodicity of around ninety years. So the power spectrum obtained by calculating the Fourier transform of a lengthy series of sunspot numbers shows two pronounced peaks (Fig. 7.2(b)). Because the cyclic variations are not precisely eleven and ninety years, the power spectrum shows relatively broad peaks which reflect the varying period from cycle to cycle. But the important feature is that the power spectrum confirms what is evident from inspecting the record of sunspot numbers – almost all the variance

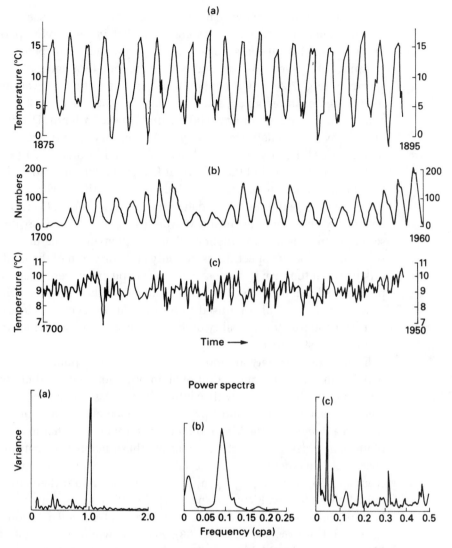

Figure 7.2 Time series and their power spectra: **(a)** the monthly temperature record for central England between January 1875 and December 1895, **(b)** the number of sunspots during the period 1700–1960, and **(c)** the annual temperature for Central England during the period 1700–1950, showing how with the increasing irregularity in the time series the power spectrum becomes more complicated (Burroughs, 1994, Fig. 2.7).

(see Section 7.3) in the last 200 years or so can be attributed to the eleven-year and ninety-year periodicities: about 65% of the variance is attributable to harmonics in the nine- to twelve-year range, while some 20% is due to the ninety-year feature. As in the case of the annual temperature cycle, the link between the time series and the power spectrum is relatively easy to see.

This direct link becomes much less obvious in the case of a typical meteorological record where we are interested in identifying periodicities in the range 2 to 100 years. Here there is no obvious cyclic behaviour in the variances of the year-to-year figures from the long-term mean. So the power spectrum (Fig. 7.2(c)) will contain a number of features of varying magnitude. In the example given, the most obvious features at 76, 23, 14, 5.2 3.1 and 2.1 years account for only about a third of the variance. Deciding which of these features is both statistically significant and of physical significance requires careful analysis, and will be considered in more detail in the next section. For the moment the important fact is that by calculating the Fourier transform of a time series, it is possible to produce the power spectrum of all the harmonics which uniquely define the observed series. Conversely, if we knew only the power spectrum it would be possible to recreate the time series by the reverse calculation. This complementary nature of the time series and its power spectrum is not only an expression of the mathematical link between the two: if the observed power spectrum were a measure of the physical behaviour of the weather and climate in the future as well as in the past, then it could also be used to forecast future events. Successful forecasting is the true test of reality of the supposed cyclic behaviour in weather and climate statistics.

There are a variety of more sophisticated computational techniques which can squeeze some additional information out of time series, but they have to be viewed with caution in the case of examining climatic change. Because, for the most part the evidence of cycles is, at best, slight, there is a danger of attaching too much importance to what may be nothing more than noise (next section). Because these techniques can present the spectra so effectively it is easy to fall into a trap. So when confronted with what look like convincing examples of cyclic behaviour there are two basic guidelines to follow. First, the spectrum should be shown in its entirety so it is possible to establish how much of the total variance is found in the principal features (see Section 7.3). Second, where spectral features are identified as being highly significant, their importance is greatly enhanced by there being an established (*a priori*) physical proposal as to why such a periodicity should occur.

7.2   NOISE

The examples presented in Fig. 7.2 provide evidence of the fact that the variability in weather and climate statistics consists of a complicated mixture of fluctuations. Leaving aside the recognisable regular variations (e.g. daily and annual cycles) and other identifiable examples of climate change (see Figs. 1.1 and 1.2), there are quasi-cyclic and apparently random fluctuations. The latter are an important aspect of analysing the nature of climate change. They are a consequence of the chaotic nature of the global weather system and are often termed *noise*. The meaning of this term does, however, warrant some closer examination.

If the fluctuations in weather and climate were truly random on every timescale the power spectrum of the fluctuations would have an equal probability of any frequency components being present. While any particular time series might exhibit a different mixture of spectral components, on average the power spectrum of such fluctuations would have equal amounts of variance in any unit interval of frequency. This distribution is often termed *white noise* – this expression is derived somewhat loosely from optical spectroscopy where *white light* contains all the visible frequencies although not in equal amounts.

The natural variability of the climate is not this simple. Many of the components of the system (see Chapter 3) have different frequency characteristics. Slowly varying factors like snow cover, polar ice, sea surface temperatures and soil moisture build in inertia and mean that the weather has a 'memory' and so is more likely to exhibit greater low-frequency fluctuations than higher frequency ones. Again in the terminology of optical spectroscopy, such noise is defined as being 'red', denoting that its distribution is weighted towards lower frequencies. In effect, because the weather has a better recollection of recent events, the short-term variations are damped out more than the longer term ones, as with the passage of time the connections become more tenuous. The theoretical distribution of red noise depends on the assumptions made about how any connections between successive events decay over time. These draw on available statistics to make an estimate of the typical time constant of the climate's memory. This construction may then be used to assess the significance of what appear to be real features in the calculated spectrum of any time series. In practice, this means that lower frequency/longer period cycles have to contain a greater proportion of the observed variance to achieve the same significance as higher frequency/shorter period feature (see Fig. 9.4 for an example of how this type of analysis is put into practice).

Fluctuations in instrumental records frequently exhibit red noise. In the case of proxy data the situation is more complicated. Because the link between the observed variable (e.g. tree-ring width) and the meteorological parameters (e.g. rainfall) is subject to uncertainty, there are additional random errors. While these errors can be reduced by the careful calibration of more recent proxy data using modern meteorological records, the problem cannot be eliminated. As a consequence, there is greater randomness in the inferred meteorological variability. This will produce white noise in any computed power spectrum. At the same time the underlying weather will have contained red noise. So spectral analysis of proxy data will contain both white and red noise; this combination is often referred to as 'pink' noise. This means that in any consideration of the significance of spectral features obtained from the analysis of proxy data an estimate of the pinkness of the background noise must be calculated.

7.3   MEASURES OF VARIABILITY AND SIGNIFICANCE

In any meteorological series the properties of the identified components of the variability can be defined in terms of the standard deviation (σ) (Box 7.1). This definition has three important applications. First, if the fluctuations are considered to be randomly distributed about the mean then observations can be made about the probability of a single extreme event, or run of extreme events occurring by chance. There is a 32% chance that any observation will be one standard deviation (σ) from the mean, and approximately a 5% chance it will occur 2σ from the mean (Fig. 7.3).

The assumption that residual variance is randomly distributed only applies in limited circumstances. Two particular criteria are important. First, the series must be stationary. This means that there is no significant long-term trend or other form of significant change in the climate during the period covered by the series. Secondly, the fluctuations are evenly distributed about the mean. This is a reasonable approximation in terms of temperature and pressure statistics. But in the case of rainfall and wind speed where the distribution of almost all the figures is markedly skewed toward low values (Fig. 7.4) because there are, in most records, far more observations of, say, zero or low rainfall than there are of very wet periods. The key to using the best statistical techniques for such series, is to establish the statistical nature of the basic distribution of the data. Only when it is clear what form the distribution about the mean takes, is it possible to decide what is the appropriate form of statistical analysis to apply to the series.

When examining instrumental records much of the variance is the product of the natural variability of the climate. This makes the detection of any trend, periodicity or sudden shift hard to detect. In addition, extreme events

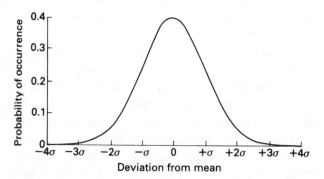

Figure 7.3 The distribution of random fluctuations in a measured variable can be represented as 'normal' curve, which shows that the probability of any particular value being observed is related to the deviation from the mean. This distribution is usually expressed in terms of the standard deviation (σ), and shows that 68% of observations will be within one standard deviation ($\pm\sigma$) of the mean, 95% will be within two standard deviations ($\pm2\sigma$) and well over 99% will be within three standard deviations ($\pm3\sigma$) (Burroughs, 1994, Fig. A.1).

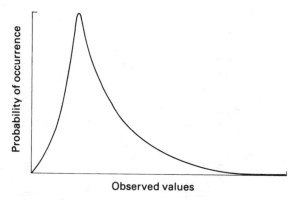

Figure 7.4 With some meteorological variables, the distribution is skewed towards lower values and the scatter cannot be expressed in terms of the deviation from the most probable value (Burroughs, 1994, Fig. A.2).

exert a substantial influence on short-term trends, and their spacing has a special fascination for statisticians.

The best way to consider these issues is to consider specific examples. In a simple case (e.g. annual figures for average temperature), where the values are approximately randomly distributed about the mean, the significance of any linear trend in the figures can be assessed in terms of the size of the rise from the beginning to the end series as compared with the standard deviation about this trend. Standard curve-fitting programs using least-squares analysis to estimate trendlines are a standard feature of PC statistical packages and provide an estimate of the significance of any trend. They can also provide higher-order polynomial fits, but without some physical explanation as to why the time series should follow such more complicated patterns, this analysis is nothing more than a statistical refinement which is unlikely to offer new insights.

Even in the case of apparently simple time series which show a fairly clear trend and a reasonably normal distribution about the mean may disguise more complicated behaviour. For instance, the CET series of annual figures (Fig. 7.5(a)) shows a distinct upward trend (0.71 °C over the period). If we use the standard analysis of a least square linear trendline the correlation coefficient (see Box 7.2), we get a value of $r = 0.32$ ($r^2 = 0.102$). While highly significant in purely statistical terms, this rise accounts for only 10% of the variance in the series. When, however, the monthly statistics are examined a more fragmented picture emerges. First, most of the temperature rise is in the winter half of the year (see Fig. 4.21) and, if the analysis is conducted on winter (December to February) temperatures, although the absolute rise in temperature (1.14 °C) is greater (Fig. 7.5(b)) the significance of the trend is rather less because the value of r is 0.24 ($r^2 = 0.06$), because of the greater variability from year to year. So, although most of the warming is concentrated in this part of the year, and the seasonal trend remains statistically significant it now accounts for only 6% of variance. This means the

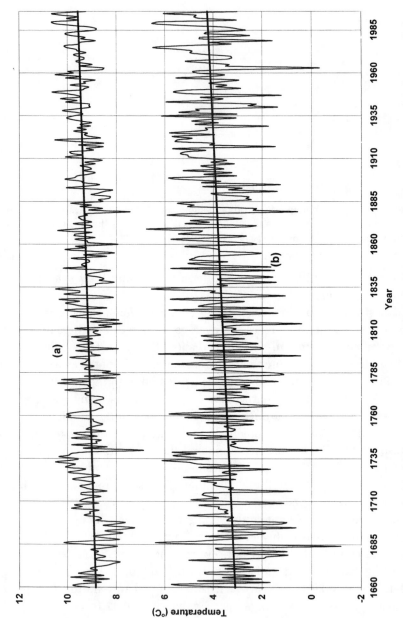

Figure 7.5 The linear trends for the temperature of central England over the period 1660–1996 for **(a)** the annual data and **(b)** the winter months (December to February) show a marked warming. In both cases this warming is significant, but although the temperature rise is greater in winter, this trend is less significant because the variance from year to year is correspondingly greater.

184

BOX 7.2 INTERPRETATION OF CORRELATION COEFFICIENTS

A fundamental aspect of statistical analysis is to establish whether there is a significant link between two variables. Here this could be how a meteorological variable changes with time, or how some weather-sensitive parameter (e.g. cereal prices, tree-ring widths or wine harvest dates) vary as a function of some meteorological variable. In the case of a meteorological time series $X(t)$ (see Box 7.1), the detection of a trend is possible by estimating whether there is a function which will fit the observed data better than the mean value $\bar{X}$. This curve could take many forms, but here we will consider only the linear trendline, which is calculated by the method of least-squares to obtain the minimum value of

$$S_m^2 = \sum_{i=0}^{i=N} (X_i - X_i')^2$$

where

$$X'(t) = a + bX(t)$$

and is the calculated trendline value of $X_i'(t)$ for the ith observation as opposed to the observed value X_i. This calculation is a complicated process, which is described in standard statistical texts (see Further Reading), and performed automatically by many statistical programs on standard PCs. The important feature of this calculation is that the minimum value of S_m is less than the same calculation using the mean value of $\bar{X}$:

$$S_0^2 = \sum_{i=0}^{i=N} (X_i - \bar{X})^2$$

where

$$\bar{X} = \frac{\sum_{n=0}^{n=N} X_n}{(N+1)}$$

To the extent S_m^2 is smaller than S_0^2 is a measure how much better a fit $X_1(t)$ is for the time series $X(t)$, as opposed to $\bar{X}$, and this is usually expressed as the *coefficient of correlation*, which is defined as:

$$r = \pm\sqrt{1 - \frac{S_m^2}{S_0^2}}$$

If the fit is good the ratio of the sums of squares will be close to zero and r will be close to ±1. If the fit is poor the ratio of the sums of the squares will be close to unity and r will be close to zero. The same type of analysis can be conducted to estimate the goodness of a fit between $(N + 1)$ meteorological observations and some proxy measurements (e.g. growing season temperature and wine harvest date – Section 6.3) for the same period.

The interpretation of a correlation coefficient (r) requires care. As a general rule, the value r can be defined as significant at a given level by a simple numerical relationship. For instance, a correlation is significant at the 95% level (i.e. there is only a 5% chance it is the product of chance) if:

$$r \leq -1.96/\sqrt{N} \quad \text{or} \quad r \geq 1.96/\sqrt{N}$$

Because in many examples of studying climatic change the value of N is large (i.e. 100 or much greater) this means that values of r of 0.2 or considerably less can indicate, say, a significant linear trend in the rise or fall of some meteorological time series. But does this have a meaning in interpreting what is really going on?

The way to interpret r is in terms of how the least-squares fit of a linear relationship to a set of data reduces the variance compared with the value obtained with the original assumption of there being no identifiable relationship. This is given by the ratio of S_m^2 to S_0^2, the value of which is $(1-r^2)$. In the case of a lengthy meteorological series where the value of r^2 is 0.1, a value of $r = \pm0.31$ is highly significant (see Section 7.2). But in terms of the variance all it is saying is that the linear trend is explaining only 10% of the variance in the series, and the remaining 90% is due to other causes. For a time series of instrumental observations, the identification of a significant trend is illuminating, even though it represents so little of the variance. For proxy data, however, where there is the additional question of whether some of the long-term variance has been filtered out by the process of data extraction (see Section 6.4.1), it may be even more difficult to attribute significance to any trend in time series.

greater fluctuations in winter temperature from year to year make it more difficult to establish the causes of variance and emphasise the importance of understanding underlying processes at work when interpreting statistics.

An additional insight into the nature of the variance of winter temperatures in England may be found in looking at the distribution of daily figures in the CET record, which have been produced by the UK Meteorological Office for the period 1772 to the present. In Fig. 7.6 the distribution of values for January for the fifty years 1772–1821 is compared with those for 1946–95. For both periods it is distinctly skewed towards higher temperatures with a long tail of low temperatures. But, although the latter period was nearly 1.4 °C warmer than the earlier one, the range of extremes is virtually unchanged. The real differences are in the central regions where, on the smoothed curves, the median has shifted 3.5 °C reflecting that there are substantially more cold days in the first period and more mild days in the latter. Analysis of December temperatures shows the same behaviour, but February has shown no appreciable warming and the distribution during the two periods is effectively identical.

A possible explanation of the shift in distribution may be found in terms of the changing incidence of winter weather regimes. If, for the British Isles,

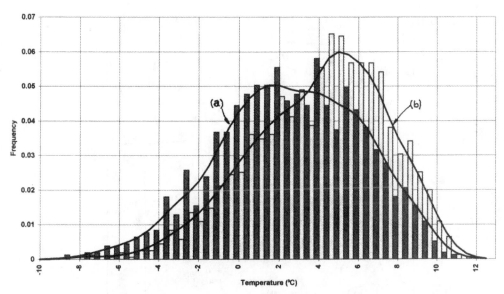

Figure 7.6 The distribution of daily temperatures for central England in January for **(a)** the period from 1772 to 1821, and **(b)** from 1946 to 1995, showing a large shift in the median temperature between the two periods but relatively little change in the extreme values (Burroughs, 1997, Fig. 5.14).

these represent, say, five broad patterns which result in different types of air masses dominating the weather in winter (e.g. continental polar [cP], maritime arctic [mA], maritime polar [mP], maritime polar westerly [mPw] and maritime tropical [mT] including rare intrusions of continental tropical air ([cT] – Fig. 7.7), then it is possible to attribute the different distributions in Fig. 7.6 to the changing incidence as shown in Fig. 7.8(a) and (b). Although this attribution is designed solely to reproduce the observed distribution, and is not linked to actual observations of how the incidence of particular weather regimes have changed, it illustrates an important physical point. This is that the oddities in the statistics become more understandable when explained in terms of a simple climatological model. In effect the substantial change in distribution without a comparable shift in the most extreme values can be interpreted in terms of a decline in easterly and northerly weather patterns, which bring cold arctic air to the British Isles, and an increase in westerly and southerly patterns bringing milder air, but no significant change in the properties of the air masses involved. So this analysis suggests that the changes affecting winter temperatures in the British Isles over the last 200 years or so are a matter of a shift in weather patterns rather than a significant warming or cooling of the northern hemisphere. This observation is also consistent with the comments made on the importance of circulation regimes in Sections 3.2, 3.6 and 8.3.

Alongside the question of the incidence of different weather regimes there is the matter of the spacing of extreme events (e.g. cold winters, hot

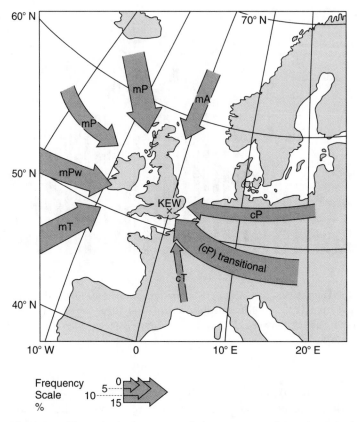

Figure 7.7 The principal sources of air masses reaching the British Isles (Barry and Chorley, 1992, Fig. 5.5).

summers or droughts, as defined in terms of being some multiple of the standard deviation from the mean). Any evidence of these events occurring in a non-random manner can be the subject of simple analysis known as the Sherman statistic (see Box 7.3). This approach is a useful way of checking quickly whether there is something interesting in the distribution of extreme events either in terms of their occurring too regularly or coming together in groups. The former can be seen as a sign of cyclic behaviour, while the latter is evidence that the climate is switching between different states from time to time. It also has the capacity to explore quasi-cyclic behaviour which is sometimes blurred out by more complete harmonic analysis.

The same care has to be exercised when interpreting the power spectrum of a time series. Assuming that the computed spectrum is not dominated by one or two features, such as the solar cycles in Fig. 7.2(b), and the variance is spread over many frequencies, great care must be taken in how much significance is attached to the most prominent of the features. The easiest way to approach the analysis of spectral variance is to consider typical examples. Figure 7.9 shows two examples of the spectral distribution for time series. In the first (Fig. 7.9(a)) the spectrum is effectively 'white' and the expected

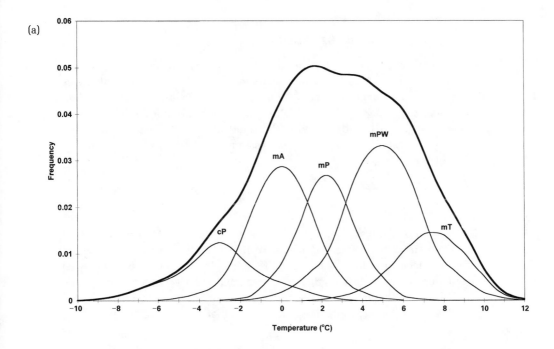

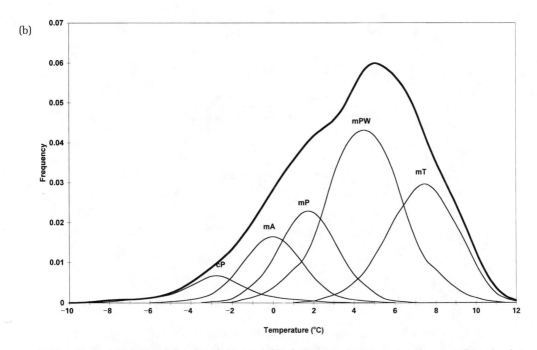

Figure 7.8 An analysis of the distribution of the daily temperature curves for central England in January (see Fig. 7.6) for **(a)** the period from 1772 to 1821, and **(b)** from 1946 to 1995 showing how they might be made up of different incidences of certain types of air masses (see Fig. 7.7) assuming that the temperature of these air masses is normally distributed (see Fig. 7.3).

BOX 7.3   SHERMAN'S STATISTIC

One useful way of examining the distribution of extreme events is to calculate what is known as Sherman's statistic. If a meteorological time series extending from time d_0 to time d_N contains n events, which exceed a certain threshold (e.g. one standard deviation above normal), and which occur at dates $d_1, d_2, \ldots d_n$, and so divide the data into $n + 1$ intervals, then the Sherman statistic is given by the expression:

$$\omega = \frac{1}{2D} \sum_{j=1}^{j=n} \left| d_j - d_{j-1} - \frac{D}{n+1} \right|$$

where D is the total length of the data (i.e. $d_N - d_0$).

The value of the statistic (ω) is a measure of the extent to which each of the actual intervals between the events differs from the average interval, and it varies according to whether the differences in time between successive events are more or less regular than expected from a random series. The probability of the value of ω being the product of chance have been calculated and are shown in Fig. B7.1. This shows that if the value of ω is high, as

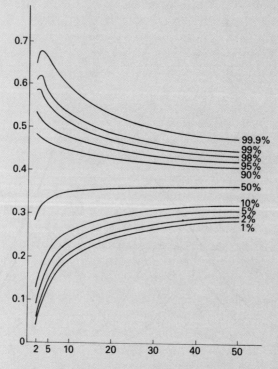

Figure B7.1 The percentiles of the distribution of Sherman's ω-statistic for values of n from 2 to 50. An evaluation can be made of the probability that large values, when events come in bunches with long spaces in between, or low values, when events are spaced regularly, may be the product of chance (after Craddock, 1968; Burroughs, 1994, Fig. A.3).

a result of the extremes coming in bunches, with large gaps in between, then this is unlikely to be the product of chance. Conversely, if the events are regularly spaced, which may be evidence of cyclic behaviour, the value of ω will be too low to be due to chance.

variance is shown as the horizontal line. This is the level for which there is an evens chance of any particular spectral component occurring. In practice, there is a considerable scatter about the expected value reflecting the random nature of the time series analysed. The mean value of the variance in the spectrum is, however, the same as the horizontal line.

In the case of a 'red' spectrum (Fig. 7.9(b)) the expected variance increases at low frequencies. The shape of this curve (see Section 7.2) reflects the fact that the series measure a variable which has a significant 'memory'. Again the actual spectrum shows considerable fluctuations about the mean variance with the most substantial features occurring at the lower frequencies, but the average variance is the same as the horizontal mean curve.

It is possible to attach statistical significance to the features in the computed spectra in Fig. 7.9. This can be done by calculating the probability of any spectral feature being more than a certain multiple of the expected variance. In many published papers (see Fig. 9.5) these levels are shown by curves indicating the levels which would be reached by chance in one-in-ten or one-in-twenty occasions, and often are labelled 90% or 95% to indicate their level of significance (i.e. with higher and higher levels there is less and less likelihood that they are the product of chance). But the important point is that for every 100 spectral components there is an evens chance that one of them will exceed the 99% significance level. In Fig. 7.9(a), the features at 0.02 and 0.375 cpa, and in Fig. 7.9(b), the feature at 0.045 cpa would most certainly fall into this category, but are almost certainly the product of chance.

Against this background, when assessing whether any particular peak is truly 'significant' it is as well to ask two questions. First, as noted in Section 7.1, has the frequency in question been predicted, in advance, as being the product of some specific physical process? This criterion is often referred to as the *a priori* requirement. Second, and directly related to identifying a physical cause, does the frequency occur in other independent time series of the same, or other meteorological variables? If the answer to both these questions is 'no' then the 'significance' of the observed feature may be nothing more than a consequence of the random nature of the time series under investigation.

The CET series of annual figures considered earlier in this section provides a good test for these principles. The Fourier transform of this series for the period 1700 to 1950 is shown in Fig. 7.2(c). As noted in Section 7.1, the six features, which are significant at the 95% confidence level, explain about

(a)

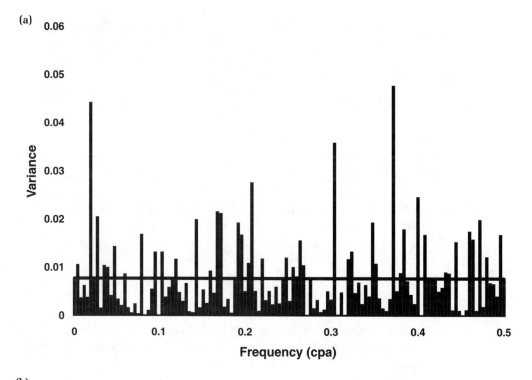

(b)

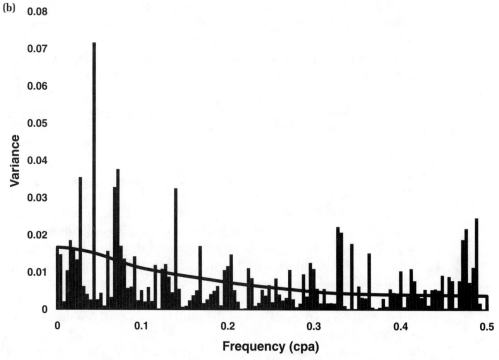

Figure 7.9 Examples of typical power spectra exhibiting **(a)** 'white' noise, and **(b)** 'red' noise. The solid bars are the spectral components and the horizontal heavy lines are the levels which would be expected on the basis of a purely random process.

a third of the variance (in this particular calculation the trend was removed before the transform was computed, as leaving it in leads to additional 'reddening' of the spectrum). Of these features, the seventy-six- and twenty-three-year peaks may conceivably be linked to solar activity (see Section 8.5), while the 2.1-year peak is possibly an echo of the QBO (see Section 3.2). The other features do not have *a priori* reason for occurring and so may be nothing more than random fluctuations.

This final examination of the CET annual figures provides a good measure of what can be extracted from a noisy time series. After identifying the trend and the six most significant features, of which only three might just be part of a periodic behaviour of the climate, we are left with nearly two-thirds of the variance apparently being a natural expression of the temperature patterns in central England.

7.4   SMOOTHING

The dominant influence of the random variations in many meteorological series means there is little purpose in going into great detail to analyse the spectral components. All that really matters is to get some sense of the underlying longer term variations by smoothing out the shorter term fluctuations. This approach helps pick out quasi-cyclic behaviour and possible significant shifts in the climate.

The simplest and most frequently used method of smoothing out a time series so that longer term fluctuations can be identified is to form a running mean of data. In its most basic form this method consists of forming the average of a given number of successive points in the time series to produce a new series. Known as the 'unweighted' running mean, this approach is widely used and easy to apply (many statistical packages associated with spread-sheets [e.g. Excel] on personal computers provide such smoothing as a standard feature). This approach does, however, have a number of limitations, which need to be considered alongside the other methods of smoothing and filtering.

To appreciate the impact of any smoothing operation on a time series we must consider how it affects the various harmonic components of the series. As we have already seen, any series can be represented by the sum of a set of harmonics. The easiest way to explain this is to take an example. If we are taking a ten-year unweighted running mean, the first obvious feature is that it will completely flatten out a ten-year periodicity of constant amplitude. This is because it will always be forming the average of one whole cycle wherever it starts from. Similarly, it will remove all the higher harmonics that are an exact number of cycles in the ten-year averaging period (i.e. 5 years, 3.33 years, 2.5 years etc.). It may also be apparent that its effect will be approximately to halve the amplitude of a twenty-year periodicity, as the ten-year running mean will take the average of half this cycle as it moves along the series.

So far, so good, but when we come to look at what it does to some of the shorter periodicities, the problems start. Take, for instance, a cycle which has a periodicity of 6.33 years (i.e. it completes 1.5 cycles each ten years). The ten-year running mean will thus form an average which contains the net effect of the additional half cycle as it moves through the series. Not only will this cycle be present in the smoothed series but it will also be inverted with respect to its original phase. It can be shown by mathematical analysis (see Further Reading), in the worst case, 22% of this harmonic passes through the smoothing process and turns up as a spurious signal completely out of phase with the original harmonic in the unsmoothed series (Fig. 7.10). This type of distortion, together with the presence of higher frequency features in different amounts, makes the use of the unweighted running mean both inefficient and potentially misleading.

To see how more efficient smoothing can be achieved, it is illuminating to consider the characteristics of an unweighted running mean in another way. The reason that high-frequency fluctuations get through is because of the way in which the smoothing deals with the data. Take, for instance, a time series of average winter temperatures which can fluctuate dramatically from year to year. These fluctuations may be random or contain some significant periodicities. The unweighted running mean is like a 'box-car' running through the series. Every data point within its span is given equal weight. So an extreme winter will enter the running mean with a sudden jump and exit in the same way. This means its effect on the smoothed series will show

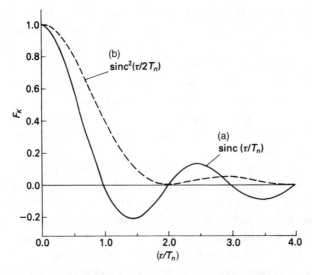

Figure 7.10 The filtering function for **(a)** an unweighted running mean and **(b)** a triangularly weighted running mean. The filtering function (F_k) is the ratio of the amplitude of the harmonics in the running mean to the amplitude of the corresponding components in the original time series. This ratio is shown as a function of the interval covered by the running mean (τ) divided by the period of the nth harmonic (T_n) in the time series (Burroughs, 1994, Fig. A.4).

up in a sharp way, even though the running mean is designed to remove all such sudden change. Given that we are only interested in the extremes to the extent that they are evidence of either longer term periodicities or a sustained shift in the climate, it would be better if each data point came into the running mean gently, built up to a maximum in the middle and faded out again. Providing this approach solves the problems of the unweighted running mean and does not introduce other distortions, it should be a better way of examining time series.

A variety of weighted running means have been explored (see Further Reading). The simplest approach would be a triangular weighting with its peak in the middle of the running mean (see curve (b) in Fig. 7.10). It turns out, however, that it is possible to design more efficient running means to act as relatively sharp 'low-pass' filters that remove virtually all the harmonics above a certain 'cut-off' frequency. The remaining harmonics are present in the series without any phase distortion, but close to the cut-off frequency their amplitude is substantially reduced (Fig. 7.11). The choice of the mathematical form of the smoothing operation is a balance between achieving a sharp cut-off and minimising both the computational effort and the number of terms needed to produce the required smoothing effect. The latter is important because in general the sharper the cut-off the larger the number of terms that have to be

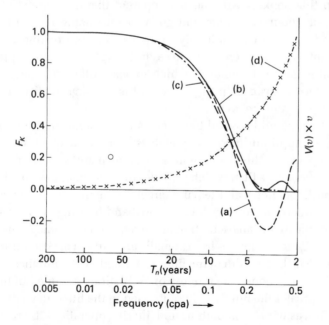

Figure 7.11 The filtering function of: **(a)** a five-year unweighted running mean; **(b)** a seven-year triangularly weighted running mean; and **(c)** an eleven-year binomially weighted running mean. The frequency scale is logarithmic, so to show the impact these running means have on 'white' noise at **(d)**, the curve of constant variance per frequency interval ($V(\nu)$) is multiplied by the frequency (ν) so that equal areas under the curve represent equal variance (Burroughs, 1994, Fig. A.5).

used. This means that the ends of the series are effectively wasted in achieving an efficient smoothing, and if there are only a limited number of observations in the series this can be a high price to pay. As a general observation, using a binomial or Gaussian weighting is a good compromise.

If the main interest is the frequency distribution, it is possible to adopt a more selective procedure. The straightforward operation of smoothing time series using either a weighted or an unweighted running mean is only a specific example of the more general technique of filtering. Instead of simply working with a 'low-pass filter' which leaves the low-frequency harmonics in the series unaltered and easier to see, there is no reason why this practice should not be extended to suppress both high- and low-frequency components and let only a limited range of frequencies through. The advantage of this process is that, unlike harmonic and spectral analysis (see Section 7.1), it permits the examination of the persistence of periodic features throughout the duration of the series. By comparison, the power spectrum is only about the mean amplitude of apparently significant oscillations while any variations in their amplitude or frequency over time are transformed into other components of the spectrum. So if these changes are appreciable, spectral analysis makes it harder to identify the real nature of the fluctuations.

This distinction is important. Wherever the question of cycles is addressed in this book, it will become apparent that convincing evidence of periodic behaviour can come and go with tantalising regularity. After several periods a cycle can suddenly disappear, only to reappear at some unspecified interval later, or shift phase and amplitude, or disappear for good. So mathematical techniques which expose different aspects of this frustrating behaviour can help to pin down the physical reality on causes of any supposed cyclic behaviour.

Ideally, a filter should pass all frequencies within a narrow band without any change in amplitude and completely suppress all other frequencies (Fig. 7.12). In practice, this is impossible to achieve and compromises have to be made in choosing a filter which provides the best combination of removing unwanted frequencies and leaving largely unaltered the frequencies of interest. The underlying approach to narrow band filtering is to construct a filter which is effectively an oscillation of the required frequency. The amplitude of this oscillation increases from a small value up to a maximum and then reduces again. The bandwidth of the filter is defined by the number of points used in the filter and hence the number of oscillations included in the computation – the greater the number of points used in the filter, the narrower its bandwidth. But, as with all smoothing and filtering operations, there is a pay-off between the narrow bandwidth of the filter and both the computational effort and the available data. In particular, if a sharp filter involves using a significant number of the available data points to compute a single point in the smoothed series, it limits the scope of the analysis. Moreover, if the data contain a considerable amount of noise, too precise a focus on a narrow frequency

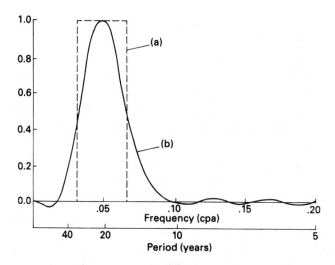

Figure 7.12 A comparison between **(a)** an ideal statistical filter which removes all unwanted periodicities and leaves unaltered those periodicities which are of interest, and **(b)** what can be achieved in practice. In **(a)** the filter transmits periods between fifteen and thirty years whereas **(b)** only transmits 20.6 years unaltered and reduces the amplitude of all other periodicities to a greater or lesser extent depending on how close they are to 20.6 years (Burroughs, 1994, Fig. 2.6).

range may serve little purpose as it will only produce a beautifully smoothed picture of the level of noise in a narrow frequency range. So, as with so many aspects of the search for order in meteorological series, there is a compromise to be struck in dealing with the limitations of the data.

7.5   MULTIDIMENSIONAL ANALYSIS

Although much of the evidence of climate change is derived from time series of a single meteorological parameter for a given site, where greater quantities of data are available it is possible to build up a spatial picture of change. This multidimensional analysis is central to unravelling the causes of climate change. For example, there is little doubt that there has been a significant warming of the global climate during the last 100 years or so (see Section 4.10). What is more difficult to establish, however, is how much of this warming is due to the natural variability of the climate and how much is due to human activities. In Chapters 9 and 10 we will discuss the various attempts to model climatic change and explain the form of rise in global temperature during the twentieth century. But, in statistical terms, we do not have an adequate knowledge of the natural variability of the climate on the timescale of decades to centuries. Since it may take a very long time to establish this background variance, a short cut is to examine whether natural variability will exhibit different spatial patterns to the impact of human activities. This requires three-dimensional statistical analysis of patterns of change, as compared with what is predicted using computer models. This analysis is often termed 'fingerprinting' (see Section 10.4).

The handling of the variation over time of two- or three-dimensional patterns of meteorological variables takes us into the realms of matrix algebra. Because this involves large amounts of data, with too many factors varying, the essence of the statistical techniques is to reduce the number of variables without discarding essential information. The most widely used technique is known as *Principal Component Analysis* (PCA). This involves computer programs which manipulate a matrix of the data to estimate how much of the variance can be attributed to each one of a set of *orthogonal* principal components. This is done by the linear transformation of the data to identify the first principal component which represents the greatest proportion of the variance in the data. The second stage of the analysis is to calculate the next component, which is uncorrelated with the first principal component, and to estimate its contribution to the variance. The value of this approach is that in many cases it homes in on the important features of change as a large part of the total variance can be attributed to a few principal components.

Perhaps the easiest way to consider how this process works is to describe a general example. Often PCA is applied to the variation over time of the geographical distribution of climate changes. This requires the analysis of, say, pressure, rainfall or temperature data for, say, a region such as north-west Europe or the USA from around 100 stations for the last 100 years or so. If these were studied station by station the overall patterns of change is particularly difficult to identify as there are as many variables as stations. With PCA what can happen is a remarkable distillation of the data. The first principal component is often a simple measure of whether conditions across the region, as a whole, were above or below normal (i.e. wetter or drier, or warmer or colder than normal) explains a considerable proportion of the variance. The second component tends to identify how much variance is due to one half of the region being above normal while the other half is below normal. This second principal component often reflects well-known features of the circulation patterns of the region. The third component may then be the variance attributable to fluctuations which are perpendicular to the changes measured in the second principal component, and this accounts for rather less of the variance. Subsequent components will locate more complicated patterns, but often over 80% of the variance has been identified with the first three components and the analysis of the higher principal components yields diminishing returns. So, in effect, most of the analysis of all the data from so many stations can be squeezed down into three variables, with relatively little loss of information.

The reason for presenting PCA in this way is that it is widely used and inevitably the product of computer analysis. So what has been done to the data is not evident to either those exploiting statistical programs or to readers of scientific papers. It is all to easy to bung large quantities of data into the computer and press the button. Out comes a string of figures defining the coefficients of a large number of principal components (*eigenvectors*) and estimates of the amount of variables attributable to each principal component (*eigenvalues*). What can be overlooked is what precisely these figures are representing. Without a physical understanding of what the data analysis is doing for you, or what lies behind the presentations of principal components in papers, there is a risk of being misled. So, unless you are prepared to really get to grips with PCA and understand the mathematics of the analysis (see Further Reading), simple guidelines to its use are twofold. First, you restrict your analysis to those first few components which can be readily defined in terms of physical features in the weather. The second is complementary to the first, in that the analysis is unlikely to be particularly illuminating unless most of the variance can be accounted for in these few components.

7.6   SUMMARY

The brief review of statistics in this chapter is designed to emphasise two essential features of dealing with data which may contain information about climate change. The first is that there is a wide variety of statistical techniques which can be applied to the large quantities of data available. But, once you get beyond relatively simple approaches to smoothing, filtering and spectral analysis, more advanced techniques should only be used with great care, because, without a thorough understanding of what can be achieved, there are real dangers of getting out of one's depth. Part of the problem is the ready availability of computing power and statistical programs which can handle large quantities of data quickly, efficiently, but without any insight as to what is the meaning of any apparent significant variations.

The identification of physical reasons for climate change is even more important in the second aspect of statistical analysis. This is the fact that most climatic series, with a few notable exceptions (e.g. the ice ages), are dominated by largely random fluctuations. This means that the series is principally noise, and if there is any identifiable physical cause for some significant change in the series, this signal is often small. Unless we have a clear understanding why the noise should take a given form, any attempt to use particularly sophisticated forms of statistical analysis to attribute particular significance to some small part of the variance must be treated with great caution. It may be reasonable to attach importance to a simple feature like the linear trend in a series. Using incredibly powerful spectral techniques, which are designed to enhance high signal-to-noise data which has been degraded for some practical reason, may, however, do no more than enable us to examine the noise in evermore excruciating detail. This is no benefit. What we want from statistics are techniques which help us to identify the physical causes of change and, better still, assist in the production of useful predictions of future change.

QUESTIONS

1   Generate a series of random numbers by taking the last two digits from a set of consecutive numbers in a telephone directory and then subject them to statistical smoothing using a simple running mean (this is most easily done using a standard PC statistical package such as Excel or Lotus 123). What does the resulting curve tell you about the problems of interpreting the smoothing of noisy data?

2   Taking an example of a meteorological time series of, say, annual temperature or rainfall statistics found in many climatology textbooks (e.g. Hulme & Barrow [1997], or Lamb [1977]) and, using a standard PC statistical package, calculate the linear trend, the variance and the correlation coefficient for the series. Repeat the analysis for the first half and the second half of the series, and then compare the three sets of results. What do the differences in the various figures tell you about the series you have examined?

FURTHER READING

A complete reference list is available at the end of the book but the following is a selection of the best books or articles to follow up particular topics within this chapter. Full details of each reference are to be found in the Bibliography.

Burroughs (1978). This paper provides a basic description of the mathematics involved in the process of smoothing time series.

Burroughs (1994). This book provides a basic guide to the statistical methods that are used to extract information from meteorological time series.

Craddock (1968). Although somewhat dated, this book provides an accessible and basic description of the statistical techniques for examining meteorological data.

Jolliffe (1986). A useful introduction to principal component analysis.

Kendall (1976). A thorough analysis of the mathematical techniques for analysing the nature and information content of time series.

CHAPTER EIGHT

THE CAUSES OF CLIMATIC CHANGE

Lucky is he who could understand the causes of things.
Virgil. 70–19 BC

By now it should be obvious that there are a lot of things which can contribute to changes in the climate. So assessing the causes of these fluctuations opens up a huge variety of physical processes. To keep this analysis manageable we must concentrate on the most obvious factors, and then focus on how these may contribute to future climatic trends. This means we will consider both short- and long-term processes in seeking explanations of natural variability. Then the analysis will be narrowed down to the causes of more rapid fluctuations when the question of the impact of human activities is brought into the debate. So, the principal objective will be to identify those aspects of climate change which provide the most insight into how the global climate may change in the foreseeable future.

Particular attention will be paid to mechanisms for climate change which have cyclic properties. This emphasis is not based on the fact that the evidence of periodic behaviour is stronger than other forms of change but because the linking of cause and effect is easier. This is a consequence of both the ability to attribute cyclic variations to a specific cause when an *a priori* reason (see Section 7.1) is identified for the periodicity to occur and then to examine a physical link between observed fluctuations and their postulated cause. So, to the extent that they can be identified, they offer the best insights into many of the forces driving climate change. In addition there is the fact that, whether or not they are real, cycles are the subject of immense and sustained speculation.

8.1   AUTOVARIANCE AND NON-LINEARITY

The description of the principal elements of the global climate in Chapter 3 introduced the concept that they can interact with one another in a complex manner to generate change. These processes are often termed *autovariance*, because they can be regarded as an internal part of the climate and so are essential to understanding the causes of change. Only when we know how the climate can fluctuate of its own accord will it be possible to separate out the impact of external influences (e.g. solar activity and astronomical tides) and human activities (e.g. the emission of greenhouse gases or deforestation). So, although these external factors will interact with the internally generated variations, it is preferable to address these issues sequentially.

Before doing so, there is, however, one further complication to consider. This is the question of the non-linear behaviour of the climate (see Box 8.1). The consequences of this phenomenon are profound. At the simplest level, it sets fundamental limits on our ability to predict the future.

BOX 8.1   IS THE CLIMATE CHAOTIC?

A chaotic system is one whose behaviour is so highly sensitive to the initial conditions from which it started that precise future prediction is not possible. Even quite simple systems can exhibit chaos under some conditions. A condition for chaotic behaviour is that the relationship between quantities which govern the motion of the system be *non-linear*, in other words a description of the relationship on a graph would be a curve rather than a straight line. Since the physical relationships governing the atmosphere are non-linear, it can be expected to show chaotic behaviour. The performance of numerical weather forecasting shows a strong dependence on the quality of the initial data, confirming the atmosphere is a chaotic system and that day-to-day forecasts are impossible more than about ten days ahead (see Section 9.2).

Although the atmosphere is chaotic, the same need not apply to the climate as a whole. For instance, the climate in any particular part of the world at any given time of the year sticks within relatively narrow limits: the temperature hardly ever rises above $-20\,°C$ at the South Pole, or falls below $20\,°C$ in Singapore. Computer models of the climate provide a reasonable description of these global patterns and their variation with the seasons. They can also be combined with what we know about the longer-term behaviour of the oceans to make useful predictions of seasonal weather in the tropics (see Section 8.2). In addition, the much longer variations associated with the ice ages can be largely explained in terms of changes in the Earth's orbital parameters (see Section 8.7). These results imply that some features of the climate are largely predictable. So, while there appear to have been circumstances when the climate has behaved in a chaotic way in the past (e.g. sudden shifts in the circulation of the North Atlantic Ocean, see Sections 4.5 and 8.3), for the most part it is not strongly chaotic.

More specifically, it underlies the capacity of complicated systems to exhibit apparently ordered but still unpredictable properties.

In considering the basic aspects of non-linear processes our starting point is how the global climate system might behave when exhibiting some regular or approximately regular fluctuations. These may be the result of internal fluctuations (e.g. the time taken for circulation of bodies of water around ocean basins) or regular external influences (e.g. solar activity of the Earth's orbital motion). The most obvious effect of non-linear systems when subjected to a variety of cyclic forces is *harmonic generation*. This means that when forced to oscillate at a given frequency it will produce higher harmonics of the fundamental frequency. Moreover, if the system is excited by two or more frequencies, it will produce sums and differences of these frequencies. Simple harmonic generation produces multiples of the fundamental frequency. The amplitude of the higher harmonics will depend on the non-linearity of the system, but will, in general, decrease rapidly with increasing frequency. The sum and difference effects are best described in terms of two frequencies (ν_1 and ν_2). Non-linear systems acted upon by two such periodic inputs will generate not only harmonics of these frequencies but a whole range of combinations given by the general expression $m\nu_1 \pm n\nu_2$, where m and n are integers. This process produces not only high harmonics but different frequencies known as *sub-harmonics*, such as $\nu_1 - \nu_2$, $2\nu_1 - \nu_2$, $\nu_1 - 2\nu_2$, and so on, which produce low-frequency oscillations in the system.

Another interesting frequency response is known as an '*entrainment*'. If a system, which has a natural self-excitation frequency ν_1, is subjected to an input of a slightly different frequency ν_2, the system may not behave in the way described above. Instead of both ν_1 and ν_2 and the frequency $\nu_1 - \nu_2$ being present, the whole system may oscillate at ν_2 with the original self-excitation oscillation effectively entrained by the imposed frequency. The range of frequencies over which this phenomenon can occur depends on the properties of the system and is known as the zone of synchronisation. A related but more unlikely effect is that in some non-linear systems it is possible either to start or stop an oscillation by starting up on an entirely different frequency. This excitation or quenching is an entirely arbitrary consequence of the system, and usually termed *asynchronous* to reflect its unpredictable nature.

These somewhat abstract concepts may seem of little relevance to the problems of climate change, but in practice they have the potential to provide useful insights. An example may help to show this. As we will see, one of the most enigmatic features of the climate is a periodic feature of about twenty years' duration. To the extent that this periodicity is accepted as real, it is variously ascribed to solar activity (the twenty-two-year double sunspot cycle), lunar tides (18.6 years) and the inherent natural variability of certain parts of the climate, notably the ocean–atmosphere interactions in the Pacific Ocean, which appear to fluctuate on a twenty-year timescale (see Section 3.7). If a natural resonance of this ocean basin happens to have a periodicity of around twenty years and both solar and lunar fluctuations on this timescale are capable of having some influence on the climate, then how they combine will be a complicated process. In particular, at different times, we could expect either of the external periodicities to dominate or, alternatively the natural frequency could take over for a while. So, over time, any of the three frequencies could be observed with varying amplitude. Switches between each periodicity could occur at random and there could well be periods when the three processes are effectively cancelled out and there are no significant fluctuations in the twenty-year range: all of which will make the record of changes hard to interpret.

One final form of behaviour worth mentioning is the differing response of systems to self-excitation. Some require only small oscillations from equilibrium to build up. This is known as 'soft-excitation'. Other systems require much greater perturbations before they will break into oscillation. This 'hard-excitation' then appears with a sudden jump. Conversely, it will exhibit hysteresis in that as the oscillation decays, the system will continue to oscillate at lower amplitude than the original threshold needed originally to get it going. This variable or erratic response of non-linear systems to both self-excitation and forced oscillation is yet another indication of the unpredictable nature of such systems.

The basic observations about the properties of simple non-linear systems have considerable implications for something as complex and non-linear as the global climate. Any propensity for parts of the system to oscillate at their own given frequencies will add or substract to one another to produce a wide variety of fluctuations whose period and amplitude will vary with time. Furthermore, the properties of entrainment and excitation mean that, if by chance, certain parts of the global climate start to oscillate at some frequency, the onset of this resonant behaviour will not be predicted and its breakdown will equally well be sudden and unexpected, irrespective of how long the apparently cyclic episode has lasted.

8.2   ATMOSPHERE–OCEAN INTERACTIONS

The consequences of non-linear phenomena are central to considering how the atmosphere and the oceans interact to produce longer term fluctuations. This interaction is, however, a classic example of the 'chicken and the egg' in that there are no obvious starting points in analysing the circular nature of the processes involved. So, it can be argued that the capacity of the atmosphere to sustain circulation patterns, which can produce changes in ocean temperatures which then reinforce these anomalous patterns, means that the atmosphere is in the driving seat. Conversely, the alternative argument is that the massive thermal inertia of the oceans suppresses the wilder fluctuations of the atmosphere and so dictates how the climate behaves over periods of a year or more.

The simple 'linear' answer to this question is that the atmosphere controls the short term, while the oceans define the longer term. This view is, in part, supported by the observed spectrum of climatic fluctuations (red noise – see Section 7.2). While this can explain some of the quasi-cyclic behaviour of the climate, it cannot, however, resolve the issue of more sudden, but lasting shifts which are so important in understanding the non-linearity of the climate.

Before considering more erratic behaviour, however, the quasi-cyclic behaviour of atmosphere–ocean interactions are best explored in terms of modelling the ENSO (see Section 3.7). The major events in the 1980s led to the development of a series of computer models which appeared to provide accurate forecasts of the progress of ENSO events a year or more ahead. The sustained event from 1991 to 1995 was not so well predicted, and the major 1997 event caught almost everyone napping until it was well in train. So the physical arguments underlying the forecasting models are being re-appraised. But, these models are able to simulate a quasi-periodic oscillation between warm and cold events, which shows they can capture some of the important features of the atmosphere–ocean interactions involved. In particular, the ability to represent the behaviour of the surface layer, so as to bring about a switch from warm to cold events or vice versa, is reassuring. This overcomes the basic problem of the model being trapped by a combination of sea surface temperatures and atmosphere circulation which maintain either warm or cold conditions permanently.

How the models achieve this important result depends on their treatment of slow-moving undulations in the thickness of the ocean surface layer, which slosh back and forth across the tropical Pacific. There are two types of wave. Close to the equator eastward moving changes in the depth of the thermocline, known as Kelvin waves (see Section 3.7), take two to three months to cross the Pacific. Westward moving changes in the depth of the thermocline are known as Rossby waves, which exhibit the same meandering properties as the long waves in the upper atmosphere (see Section 3.2). Because of the different physical properties of the ocean they are, however, much slower moving. On the equator they can take as little as three months to travel westwards across the Pacific, but, they, unlike Kelvin waves, are affected by the Coriolis force and are much slower moving at higher latitudes (at 30° N and S they take ten years to cross the Pacific). When both these types of waves reach the edges of the Pacific they tend to be reflected back in the direction they have come from, but they switch type, so Kelvin waves make the return as Rossby waves and vice versa. Computer models of these processes, when combined with realistic representations of atmospheric circulation patterns, exhibit the property of switching back and forth between El Niño and La Niña conditions every three to five years.

The other feature of some of the models is that they are more likely to produce quasi-cyclic oscillations after there has been a significant perturbation
of conditions in the equatorial Pacific. But, when any oscillations die
down the system becomes more chaotic. This response is precisely
what would be expected from the study of simpler non-linear systems in
terms of those which exhibit the property of requiring hard excitation before
oscillating. Furthermore, it reflects the experience of the early 1990s
when less extreme El Niño (warm) conditions prevailed for nearly five
years, whereas the models repeatedly predicted an earlier switch to colder
conditions.

These problems were compounded by the difficulties experienced with
forecasting the extreme event in 1997. One component of these problems
was the part played by what are known as intraseasonal variations in the
tropical weather. These are pulses of strong winds and rain that travel eastward around the equator in about thirty to sixty days. Known as the
Madden–Julian Oscillation (MJO), after the scientists who first identified
it, this phenomenon is at its strongest from December to May, when it is
one of the most intense weather systems in the tropics, pumping huge
amounts of heat into the atmosphere. More relevant is the fact that if one
of these bouts of activity, with its strong convection and westerly winds
happens to hit the western Pacific just as an ENSO warming event is
about to break, it can stimulate its rapid development. This is what appears
to have happened in early 1997. It is, however, in the nature of these MJO
oscillations that they are difficult to forecast, being as chaotic as other
aspects of atmospheric circulation. They may also have been instrumental
in the sudden onset of La Niña in 1998. So, it appears that with the ENSO, at
some times, the chaotic nature of the atmosphere may be in the driving seat
but, like the proverbial supertanker, once the ocean is heading in a given
direction it takes a long time to turn it around.

The quasi-periodic variations of the NAO (see Section 3.5) may also be
the result of a similar set of feedback processes across the North Atlantic and
the adjacent continents, and also with wider global atmosphere–ocean interactions. There is evidence that the position of the Gulf Stream is linked to
the NAO. Measurements of the 'north wall' of the current between 1966 and
1996 show that 60% of the variance of the position could be predicted in
terms of the NAO. Moreover, much of the remaining variance could be
linked with fluctuations in the Southern Oscillation.

One possible explanation for these apparent links rests on the fact
that during the positive phase of the NAO the cooling in the Labrador
Sea leads to creation of more deep water. This flows southwards down
the east coast of North America where it could influence the strength and
direction of the Gulf Stream. This, in turn, could lead to changes in the
temperature of the eastern North Atlantic which in due course could tip
the NAO back into its negative phase. But, so far, no convincing measurements have yet been obtained which substantiate this explanation of why

the NAO switches back and forth, or why it should do so at any particular frequency.

8.3 OCEAN CURRENTS

The major part played by ocean currents in the transport of energy to high latitudes (see Section 3.6) means that any changes in this pattern could have substantial climatic implications. In addition, it is apparent that different ocean circulation patterns associated with earlier distributions of the continents were a major factor in the radically different climatic patterns that existed for much of the geological past (see Section 4.1). What is more, changes during and around the end of the last ice age (see Section 4.5) suggest that the circulation in the North Atlantic can undergo sudden and substantial shifts. These could lead to the global climate being able to exist in distinctly different regimes even though the overall energy balance of the system had not changed appreciably. So understanding how the large-scale motions of the oceans can shift as a result of both their own natural variability and external perturbations is central to unravelling the causes of climate change.

While there is no doubt about the importance of different patterns of ocean circulation in establishing past climatic regimes, the real issue is whether sudden changes are relevant to the current debate on the impact of human activities. This revolves around the question of how sensitive the Great Ocean Conveyor (GOC) is to changes in the amount of freshwater entering the northern North Atlantic. Modelling work suggests that circulation patterns are extremely sensitive to run-off from the continents, the number of icebergs calved off Greenland and the amount of precipitation from low pressure systems tracking north-eastwards past Iceland and into the Norwegian Sea. Small changes in the total input may be able to trigger sudden switches to alternative patterns which carry warm surface water less far north before it sinks and returns southwards. This pattern would reduce sea surface temperatures around southern Greenland and Iceland by 5 °C or more. This would have a drastic impact on the climate of Europe and completely alter the atmospheric circulation patterns of the northern hemisphere.

The real issue is whether the changes produced in the models simply by altering the balance between evaporation and precipitation across the northern North Atlantic are climatically realistic. While the behaviour of the model can be tuned to produce chaotic shifts in circulation, the fact that, with one short-lived exception (see Section 4.5), such large shifts have not occurred in the last 10,000 years can be seen as evidence that bigger perturbations are required to get the GOC to switch to a different mode. One explanation is that changes seen both at the end of the last ice age and during its various fluctuations were the result of the total or partial collapse of the ice sheet over North America which produced a surge of icebergs to flood out into the North Atlantic. These Heinrich events, which can be seen in the records of ocean sediments (see Fig. 4.13), would have a much greater impact on the GOC. So, at the moment the potential for the current climate to undergo such sudden shifts is an unresolved issue. But, the fact that human activities may be leading to greater and more rapid changes than anything seen in the last 10,000 years may be sufficient reason to worry about the potential of the GOC to 'flip' into a different mode (see Section 10.5).

The longer term effect of different ocean circulation regimes is bound up with the question of continental drift. As has been noted in Chapter 4, the changes that occurred not only in the distribution of the continents, but also in the seaways that opened and closed at various times (e.g. the opening of the Drake Passage around 25 to 30 Ma, or the closing of the Panama Isthmus around 3 Ma) had a profound effect on ocean circulation and hence global climate. In terms of causes of climate change, it is, however, a moot point as to whether these changes can be regarded as principally matters of ocean circulation. What is beyond doubt is that the different circulation regimes that existed before and after these tectonic developments were capable of maintaining radically different climatic states.

8.4   VOLCANOES

It was Benjamin Franklin who first identified the potential of volcanoes to alter the climate. He suggested that the bitter winter of 1783–4 in northern Europe was caused by the dust cloud produced by the huge eruption of Laki in Iceland in July 1783, which dimmed the sun in Paris for months on end. This eruption, and others at low latitudes (e.g. Tambora) are plain to see in the Greenland ice cores (see Fig. 6.13).

Explosive volcanic eruptions can inject vast amounts of dust and, more significantly, sulphur dioxide into the upper atmosphere where this gas is converted into sulphuric acid aerosols. At altitudes of 15 to 30 km, where there is no significant vertical motion, these minute particles remain suspended for up to several years and are spread round the entire globe. A dust veil in the upper atmosphere absorbs sunlight. This heats the stratosphere but causes compensating cooling at lower levels, as less solar radiation reaches the Earth's surface. Analysis of past eruptions suggested that these physical processes did have a significant impact on the climate. There was, however, considerable doubt about just how big the impact on the global climate was. This uncertainty was the product of the fact that any cooling is accompanied by shifts in global weather patterns. Because the analysis, prior to the late nineteenth century, was based largely on observations of climate in middle latitudes of the northern hemisphere, parts of which experience disproportionate cooling, it was hard to be certain that observed changes were representative of global changes. These problems were compounded by the fact that following Krakatau in 1883 there was no truly significant eruption until Agung in Bali in 1963. What was missing was an adequate global picture of the direct impact of a major eruption on the temperature of the atmosphere.

In the 1980s things changed. First, there was the eruption of Mount St. Helens in 1980. Contrary to popular belief, however, this had no appreciable impact on the climate because it blew out sideways it did not inject much of its dust high into the stratosphere. More significantly, its plume was low in sulphur compounds. So, the limited cooling effect provided climatologists with valuable insights into the essential role of sulphur in altering the climate. The eruption of El Chichón in Mexico in 1982, although a relatively small volcano, its emissions were very high in sulphur. Satellite measurements of the resultant stratospheric aerosol cloud provided confirmation of the climatic importance of sulphur. The massive eruption of Mount Pinatubo in the Philippines in 1991 offered a second chance to test these hypotheses (Fig. 8.1). It injected around twenty million tonnes of sulphur compounds into the stratosphere, and was by far the most climatically important eruption this century.

The eruptions of El Chichón and Pinatubo have been used to test theories of their impact on the climate, both by direct measurements and by the use of computer models of the climate. They are particularly valuable as the scale of the maximum perturbation caused by Pinatubo was estimated to be equivalent to a global reduction in solar energy reaching the Earth's surface of 3 to 4 W m^{-2}. This change is of the same scale as the radiative forcing due to the equivalent of doubling of CO_2 in the atmosphere

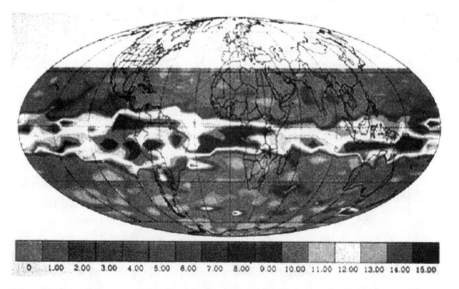

Figure 8.1 Satellite measurements of the spread of the dust cloud from the eruption of Pinatubo in June 1991, showing that within five weeks the dust girdled the globe (Burroughs, 1997, Fig. 5.11).

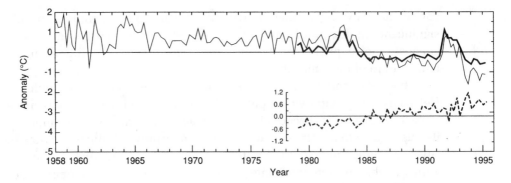

Figure 8.2 The cooling of the stratosphere between altitudes of 16 and 21 km observed by radiosondes (thin line) and satellites (thick line), together with the difference between the two sets of observations (dashed line) (IPCC, 1995, Fig. 3.7).

(4 W m^{-2} – see Section 2.1.3). Accurate measurements of the temperature of the stratosphere (Fig. 8.2) using microwave radiometers on weather satellites (see Section 6.2) clearly show how the dust clouds from these eruptions warmed the upper atmosphere. The corresponding cooling of the lower atmosphere was not only observed in the surface temperature record, but also accurately predicted by computer models of the climate (see Section 9.2).

What these observations confirm is that major volcanoes do cool the climate at ground level and there is a compensating and somewhat greater warming of the stratosphere. The effects of a single eruption lasts for around two to three years. This supports Benjamin Franklin's original hypothesis and the widespread assumption that the massive eruption of Tambora in Indonesia in 1815, which injected five to ten times more material into the stratosphere than Pinatubo, was responsible, in 1816, for the 'year without a summer' when exceptionally late frosts destroyed crops in New England, and French vineyards experienced their latest wine harvest in at least the last five centuries. This cooling may also have been a contributory factor towards the consistently low temperatures in the 1810s (see Section 4.9). But Tambora is, at most, only part of the story, as the cooling of this decade was well and truly underway before the eruption occurred, and it has been postulated that an earlier, as yet unidentified, eruption in 1809, which shows up in Antarctic ice cores, may have set the cooling in motion.

This uncertainty about when and where earlier eruptions took place is a major limitation in establishing the role of volcanoes in past climate change. Recent studies of tree rings at mid- to high-latitudes around the northern hemisphere have been combined with ice-core data to produce an improved analysis of the impact of eruptions on the climate. This confirms that major eruptions produce a substantial drop in the summer temperatures for two or three years. If anything, the effect on winter temperatures is a warming, owing to stronger westerly circulation at mid- to high-latitudes. This analysis has identified 1601 as the coldest summer in the last 600 years, possibly owing to the eruption of a volcano in southern Peru the year before: 1816 is the second coldest year in this period.

As for longer term effects on the climate, the short-lived impact of volcanoes suggests that they can only trigger lasting change if they coincide with other perturbations which are stimulating cooling as well. An interesting example of this is the fact that the largest eruption in the last million years – Toba in Sumatra, 73,000 years ago, which was at least five times the size of Tambora – coincided with a rapid cooling during the development of the last ice age (see Table 4.2 and Fig. 4.14). This eruption could have produced sufficient cooling for perennial snow cover to form for a number of years in, say, northern Canada. At a time when the Earth was already slipping into a cold phase of the last ice age, this snow cover with the additional cooling effect of reflecting more sunlight into space may well have tipped the balance.

An alternative explanation for these coincidences is that volcanic eruptions could be triggered by sudden sustained changes in atmospheric circulation. These alter the stress on the Earth's crust and are even detected in tiny changes in the length of the day (i.e. how fast the Earth rotates). Another mechanism which is supported by the timing of some major eruptions is that changes in sea level alter the crustal loading in the vicinity of volcanoes close to the shoreline and this triggers some eruptions. So, as in many other proposals for causes of climate change, an appreciation of the complex web of feedback mechanisms between cause and effect are central to forming a balanced understanding of the processes at work.

8.5   SUNSPOTS AND SOLAR ACTIVITY

As noted in Section 2.2, sunspot activity has been monitored since the seventeenth century. The average number of spots and their mean area fluctuate over time in a more or less regular manner with a mean period of about 11.2 years (see Fig. 2.10). During this fluctuation the rate of increase in their number exceeds the rate of decrease, the period varies between 7.5 and 16 years, and the amplitude varies by about ±50%. The variation in number during each period is more than two orders of magnitude greater than for any shorter period. It ranges from virtually no spots during the minimum in solar activity to just over 200 in the most active cycle which peaked in 1957 (see Fig. 2.10). Each cycle begins when the spots show up in both the northern and southern hemispheres some 35° away from the solar equator. As the cycle develops, the older spots fade away and new, more numerous spots appear at lower latitudes (Fig. 8.3). Towards the end of each cycle the number decreases and the spots are concentrated at latitudes some 5° from the equator. This cycle of activity does not necessarily fall to nothing at the minima, because a new cycle will start at high latitudes before the old one has died away at low latitudes. This overlap can exceed two years.

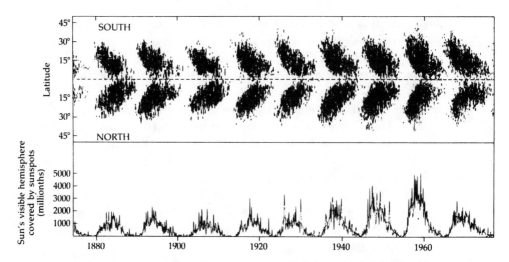

Figure 8.3 The observed variations of the number of sunspots between 1875 and 1975 together with the 'Maunder butterfly' diagram showing the distribution of sunspots with heliographic latitude and their general movement in each hemisphere during successive sunspot cycles (Giovanelli, 1984; data supplied by SERC, RGO; Burroughs, 1994, Fig. 6.2).

Closely associated with sunspots, and most easily observed near the solar limb, are brighter areas called 'faculae'. The way in which their output is linked with the incidence of sunspots has recently become an important factor in explaining how changes in solar activity could affect the weather.

Ground-based observations of the amount of energy reaching the Earth from the Sun as a function of the number of sunspots were unable to provide convincing evidence of an effect, because perturbations due to variations in atmospheric absorption swamped any small changes in the Sun's output. Satellite-borne instruments have, however, changed the situation. Starting with an instrument on Nimbus 7, launched in 1978, and followed by the Solar Maximum Mission (SMM) in 1980, a series of measurements have produced unequivocal observations of how the Sun's output varies with the eleven-year cycle in solar activity (see Fig. 2.11). While there are slight differences in the absolute energy measurement there is no doubt that the total radiance from the Sun varies with sunspot number during the twenty-first sunspot cycle from 1980 to 1992 by about 0.1% between the peak levels in 1980 and 1991 and the minimum in 1986.

The observed variations in solar energy output provide the basis for a physical explanation for linking changes in the weather and solar activity. But, the fact that solar output rises with sunspot number is a complication. Because sunspots have a lower radiative temperature it was argued that they would block some of the Sun's output and so reduce the overall energy flux. So output should decline with the sunspot number. The reverse suggests that an associated increase in the brightness of the faculae is the dominant physical factor.

A model which combines sunspot numbers with recent satellite ultra-violet measurements and ground-based microwave observations appears to provide the solution. Although the satellite only started around 1980, the microwave observations have been made since 1954 and provide a measure of the radiative temperature of faculae. In this model, proposed by Judith Lean and P. Foukal, sunspot number and faculae brightness based on microwave observations are combined to obtain a good fit with satellite observations since 1980. This model successfully explains why the Sun is consistently brighter around sunspot maxima. But the relationship between sunspot number and faculae brightness is not simple as periods of highest sunspot frequency are not necessarily those when the Sun is radiating the most energy. For instance, the model predicts that the Sun's radiance was higher at the peak in solar activity in 1980 than it was during 1959 when the sunspot number was higher.

Using this approach it has been possible to use the sunspot figures to estimate changes in solar output back to 1874. This shows that these variations were negligible between 1874 and 1945. Since then there has been a gradual increase in irradiance and a significant cyclic variation in output (Fig. 8.4). The direct link between sunspot number and solar output fits with the hypothesis that the cold period known as the Little Ice Age (see Section 4.9), and more particularly the colder weather of the late seventeenth century, was a result of an almost complete absence of sunspots (known as the *Maunder minimum*) during this period.

How the Sun's radiance varies with sunspot number is not enough to explain the connection between solar activity and the climate. The small change in irradiance (about 0.1%) during the solar cycle is an order of magnitude too small to produce the observed rise in temperature since the

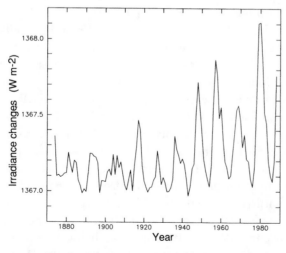

Figure 8.4 Reconstructed solar irradiance W m^{-2} from 1874 to 1988 estimated by Foukal and Lean. (N.B. The climatic effect will be only 0.175 times the calculated irradiance variation owing to area and albedo effects.) (IPCC, 1990, Fig. 2.5.)

late nineteenth century (see Fig. 4.23). So, if solar activity has played a significant part in the recent global warming, it must be amplified in the Earth's atmosphere and hence modulate the weather in a non-linear way. One possible explanation lies in the wavelength dependence of solar fluctuations. Satellite observations show that much of the flux change is concentrated in the ultraviolet (UV). For example, although only 1% of the Sun's radiation is emitted in the 200 to 300 nanometre (nm) waveband, some 20% of the observed change in solar output occurred in this range. This is not surprising – short wavelength radiation is likely to be linked with faculae brightness because faculae are regions of intense magnetic activity and have been identified as the dominant sources of high energy UV, X-rays and γ-rays from the Sun.

The climatic impact of changes in solar radiance in the UV region is an interesting proposition. Because wavelengths between 200 and 300 nm are absorbed high in the stratosphere by oxygen and ozone molecules (see Box 2.2), they initiate photochemical reactions which influence the weather at lower levels. Computer models of how the amount of solar radiation entering the lower atmosphere varies with solar activity, as a result of alterations in stratospheric ozone concentrations caused by the changing UV flux, suggest there is an amplification process at work. This reduces the amount of solar energy reaching the lower atmosphere in middle and high latitudes in winter when solar activity is high. These changes could have a significant impact on global circulation as preliminary calculations suggest that increased solar UV radiance in the lower tropical stratosphere will expand the Hadley circulation (see Section 3.1) leading to a poleward shift of the sub-tropical westerly jet and the mid-latitude storm track.

Another photochemical consequence of changing UV fluxes reaching the lower atmosphere is to affect the formation of free-radicals in the lower atmosphere, notably the hydroxyl radical (OH). This alters the production of condensation nuclei and hence the formation of clouds. In effect, more UV radiation reaching the troposphere will increase the concentration of condensation nuclei and hence make it cloudier – a process which would amplify any fluctuations due to varying solar activity.

Magnetic fields associated with the sunspot cycle could also affect the climate. The polarity of sunspots alternates between positive and negative in successive eleven-year cycles. Sunspots tend to travel in pairs or groups of opposite polarity as if they are the ends of a horseshoe magnet poking through the surface of the Sun. During one eleven-year cycle, as the spots traverse the face of the Sun in an east–west direction, the leading spots in each group in the northern hemisphere will generally have positive polarity while the trailing spots will be negative. In the southern hemisphere the reverse situation occurs with the leading spots being negative. It is this pattern that reverses in successive cycles. Known as the Hale magnetic cycle, this twenty-two-year cycle could be the key to the amplification process, as it determines the solar-induced interplanetary magnetic field direction, which is one of the controlling factors in the solar wind interaction with the Earth's magnetosphere. As a general observation the twenty- to twenty-two-year cycle has been more prevalent in climatic data than the more obvious eleven-year sunspot cycle. Of particular interest is that it is a significant feature in the global marine air temperature record (see Section 6.1). So any magnetic process which amplifies the impact of solar variability on the weather could have played a part in the observed recent global warming. One possibility is that magnetic fields affect the quantity of energetic particles emitted from the Sun. A second is that they alter the Earth's magnetic field and so influence the amount of cosmic rays (energetic particles from both the Sun and from elsewhere in the universe) that are funnelled down into the atmosphere. These effects may alter the properties of the upper atmosphere and so could affect the weather in a variety of ways, including photochemical routes discussed above.

Another way a changing flux in cosmic rays could affect the climate has been identified by Ralph Markson at the Massachusetts Institute of Technology. He proposed that the modulation of cosmic rays by the Sun leads to changes in the Earth's electric field and hence thunderstorm activity. This mechanism has three attractions. First, it requires no significant change in solar energy to alter the state of the Earth's magnetic field and stratospheric conductivity, while offering the possibility of releasing and redistributing large amounts of energy already present in the troposphere. Secondly, it does not require strong links between the upper and lower atmosphere as the electric field variations encompass the whole atmosphere from the ionosphere to the Earth's surface. Thirdly, the response of the electrical field to change in the magnetic field is almost instantaneous and this can explain how the weather responds within a day or to certain changes in solar activity.

Markson postulates that worldwide thunderstorms play a central role in maintaining a global electric circuit. So changing of the conductivity of the upper atmosphere can alter the incidence of thunderstorm activity. Greater stratospheric ionisation could lead to increased thunderstorm electrification either locally or as a consequence of global changes in the Earth's electric field, and this may alter thunderstorm development. Changes in the Sun's magnetic field will also alter the number of energetic particles it emits. This will have complicated effects on the Earth's atmosphere, which could include affecting the cloudiness caused by thunderstorms. At high latitudes the effects of the flux of solar protons, which is directly related to solar activity, will predominate. At low latitudes the magnetic field variations will modulate galactic cosmic rays and produce an effect which is out of phase with solar activity. This may explain why there is evidence of a high positive correlation between the sunspot cycle and high-latitude thunderstorm activity whereas at low latitudes it is either non-existent or negative.

Thus far we have only considered the basic eleven-year sunspot cycle and the double Hale cycle. As is obvious from Fig. 2.10, the variation in the size of the peaks in successive cycles also shows evidence of a periodicity of around 90–100 years – often termed the Gleissberg cycle. This cycle is also associated with the change in the period between successive peaks of solar activity, which lengthens as the peak levels decline. When sunspot series is subjected to detailed spectral analysis two important features emerge. First, the principal feature at around eleven years (Fig. 8.5) is split into two main peaks at 11.1 and 10.0 years. Second, about 20% of the total variance in sunspot numbers is associated with the ninety-year periodicity, which may be a difference frequency between the two features at 10 and 11.1 years. In addition, there is considerable evidence of a 200-year cycle in sunspot activity, including human observations, notably from China.

The importance of the longer cycles in solar activity is the marked parallelism between changes in global temperatures and both sunspot numbers and the length of the eleven-year sunspot cycle. Recent work by Judith Lean and colleagues, building on the model described earlier, have reconstructed variations in solar UV radiation since the beginning of the seventeenth century, and its implications for climatic change. They conclude that the correlation between changes in solar UV and northern hemisphere temperatures is 0.86 and solar forcing has been responsible for half of the warming between 1860 and 1970, and one third of the rapid increase between 1970 and 1990. In statistical terms, the reconstruction accounts for 74% of the variance in northern hemisphere temperature in the period 1610 to 1800 and 56% from 1800 to 1990. These figures are highly significant (see Box 7.2).

In considering the possible impact of longer term variations of solar activity, the best source of information is tree-ring data. Because the strength of the

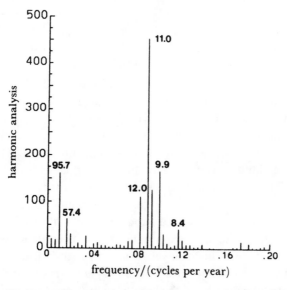

Figure 8.5 The power spectrum of sunspot numbers for the period 1700–1986 (Pecker and Runcorn, 1990, with permission of the Royal Society; Burroughs, 1994, Fig. 6.5(a)).

solar magnetic field modulates the flux of cosmic rays entering the Earth's atmosphere, it affects the production of carbon-14 (^{14}C) (see Section 6.6). When solar activity is high, the magnetic shield is strong and the amount of ^{14}C formed, and hence incorporated in tree rings, is relatively low. Conversely, at times of low solar activity, the shield is weak and more ^{14}C is formed. By measuring the amount of ^{14}C in tree ring series, and comparing it with what would be present if it had been produced at a constant rate, it is possible to build up a measure of past solar activity (see Box 6.3).

Studies of the spectra of ^{14}C fluctuations of tree-ring records going back some 9,000 years may provide evidence of periodic fluctuations in solar activity. Studies have been conducted on tree-ring data from the White Mountains in California, and from fossilised oaks in Europe (see Fig. B6.1, in Box 6.3). This analysis identified five strong periodic features in the data. These occurred at about 2,300, 500, 355, 204 and 154 years. The fact that these fluctuations have almost exactly the same periodicities and power densities, and are present in tree rings from widely separated parts of the globe suggests they are due to real physical effects. What is not clear is the extent to which they are a product of solar variability as opposed to other factors which could influence the amount of radiocarbon in the atmosphere. Alternative explanations include changes in the magnetic field of the Earth, or changes in the size and exchange rates of the carbon reservoirs on the surface of the Earth, including the biosphere and the oceans.

The existence of the periodicity at around 200 years is intriguing, as it is close to observed fluctuations in sunspot numbers. Over the last 1,000 years there appear to have been major lulls in solar activity at around AD 1280, AD 1480 and AD 1680 (the Maunder Minimum). These periods of low solar activity appear to coincide with periods of a cooler climate in the northern hemisphere, notably the period at the end of the seventeenth century (see Section 4.9). Since 1700 this pattern has been less evident, with the most marked minimum in the 1810s, and a broad reduction in the late nineteenth century (see Fig. 2.10).

8.6   TIDAL FORCES

The obvious link between tidal forces and the climate is through the direct alteration of the movement of the atmosphere, the oceans, and even the Earth's crust. The nature of these links varies in complexity. Tidal effects in the atmosphere are relatively predictable and measurable, but tiny compared with normal atmospheric fluctuations. In the oceans the broad effects can be calculated, but estimating changes in the major currents is far more difficult. When it comes to the movements of the Earth's crust, the problems are compounded by possible links with solar activity. The direct influence of the tides could influence the release of tectonic energy in the form of volcanism. Since there is evidence that major volcanic eruptions have triggered periods of climatic cooling (see Section 8.4), this would enable small extra-terrestrial effects to be amplified to produce more significant climatic fluctuations. In addition, there is evidence of intense bursts of solar activity interacting with the Earth's magnetic field to produce measurable changes in the length of the day. Such sudden tiny changes in the rate of rotation of the Earth could also trigger the release of tectonic activity (see Section 5.1).

The gravitational forces acting on the Earth as it orbits the Sun can be divided into four categories. First, there are the tides resulting from the combined pull of the Moon and the Sun. These tidal forces affect the movement of both the atmosphere and the oceans and also exert stress on the Earth's crust. Secondly, the forces exerted on the Earth by the changing positions of the other planets will play a similar but much smaller role. Thirdly, there is the possibility of the same tidal forces due to planetary motions affecting the Sun's circulation and with it solar activity. Finally, there are the orbital effects of these motions. Because these can cause the Earth to speed up and slow down in its orbit and also lead to small movements of the Sun about the centre of mass of the solar system, there is the potential for small periodic influences on the Earth's climate.

Clearly, all these tidal effects are interlinked. But, as a first step, we need to consider each potential influence separately before trying to make observations about what their combined effects may be. So, the obvious place to start is with the semi-diurnal tides in the atmosphere and the oceans. These are caused by the gravitational attraction of both the Sun and the Moon. On the nearside of the Earth the atmosphere and oceans are attracted towards both bodies, as is the Earth itself, which pulls away from its fluid envelope on the far side. Because of the Earth's rotation and the Moon's orbital motion, any particular place on the Earth's surface experiences two complete cycles of high and low tidal stress every twenty-five hours. On average the Sun's pull is roughly half that of the Moon. This means that there is a threefold variation between when the Sun and the Moon are on the same side of the Earth and pulling together, and when they are on opposite sides so that their effects are partially cancelled out.

If the Moon completed an exact number of orbits of the Earth in the time it takes the Earth to complete a circuit of the Sun, the pattern of tidal forces would be relatively simple and reproducible. But this is not the case. The nearest it comes to this is that thirteen tidal months amount to 355 days, which is sometimes referred to as the 'tidal year'. So although the Earth and the Moon return to approximately the same position after about a year, it takes much longer for more precise repetitions of alignment to occur.

This brings into play the relative motion of the Earth's perihelion – the point on its orbit when it is closest to the Sun – and the Moon's perigee – the point on its orbit when it is closest to the Earth. While the changing distance between these three bodies will exert a continual influence on the tidal forces, the relative position of the perihelion and the perigee play an important part in the longer term periodicities that may influence the weather.

There are two periods of particular interest. The first is the 8.85-year period in the advance of the longitude of the Moon's perigee which determines the times of the alignment of the perigee with the Earth's perihelion (Fig. 8.6). The second is the 18.61-year period in the regression of the longitude of the node – the line joining the points where the Moon's orbit crosses the ecliptic. This period defines the exactness of the alignment of the Moon's perigee and the Earth's perihelion.

The 18.61-year cycle is the most widely studied aspect of the tidal stress. This is because the regression of the node defines how the angle of the Moon's orbit to the Earth's equatorial plane combines with or partially cancels out the tilt of the Earth's axis. This has the effect of altering the variation in the maximum declination of the Moon from the ecliptic. Because of the tilt of the Earth's axis, the equatorial plane is at an angle of $23°\ 27'$ from the ecliptic. This combines with the angle of the Moon's orbit to the equatorial plane so that the maximum declination to the ecliptic is $28°\ 40'$ N and S. But this extreme value occurs only every 18.6 years; at the opposite extreme, halfway between when the maximum declination is the difference between

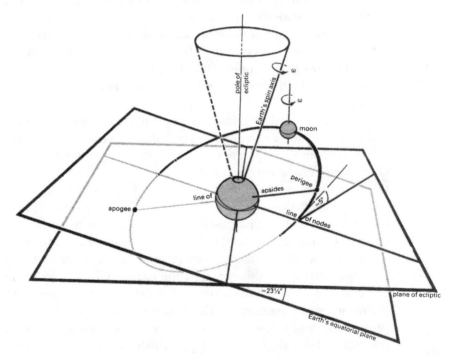

Figure 8.6 The orbital geometry of the Earth–Moon system. The Earth's average orbital motion around the Sun defines the ecliptic and the Earth's spin axis rotates about the pole of the ecliptic once every 26,000 years because of the precession. The Earth's equinox, the intersection of the equator and the ecliptic, moves along the ecliptic at the same rate. The lunar orbit intercepts the ecliptic along the line of the nodes which moves around the ecliptic because of the solar attraction. For the same reason the line of the apsides precesses in the orbital plane. The Moon's spin axis remains normal to the ecliptic (*Cambridge Encyclopaedia of Earth Sciences*; Burroughs, 1994, Fig. 6.7).

the tilt of the Earth's axis and the angle of the Moon's orbit from the equatorial plane, the value ranges only between 18° 21′ N and S. The significance of this variation is that when the declination is greatest, the tidal forces at high latitudes are greatest. Recent peaks in these forces have occurred in 1913, 1931, 1950, 1969 and mid-1988.

The importance of the 8.85-year cycle in the alignment of the Moon's perigee and the Earth's perihelion is not its direct impact on tidal forces but how it combines with the 18.61-year cycle. Calculations of tidal stress at high latitudes since AD 1100 show that there is a tidal resonance of about 179.3 years. The significance of this period is that it is yet another candidate for the general group of periodicities that have been identified as being around 180 to 200 years.

Looking beyond the tidal effects of the Sun, Moon and Earth, there is the influence of the other planets in the solar system. Of these, the motions of the giant planets Jupiter, Saturn, Uranus and Neptune are potentially the most interesting. These have orbital periods of 11.86, 29.5, 84 and 165 years respectively. So, singly or in combination they could influence the tidal forces on the Earth. In practice, because of its mass (318 times that of the Earth) and its relative proximity to the Earth, Jupiter is by far and away the most important factor in these planetary tidal forces. Moreover, the fact that its orbital period is close to the eleven-year cycle observed in the sunspot number means that it could either be a confusing factor in identifying the cause of periodicities in the weather or be directly linked with the sunspot cycle itself.

In respect of the tidal influences of the planets on the Earth's atmosphere and oceans, detailed calculations show that the scale of these perturbations are small compared with the variation of the tidal forces due to the motions of the Earth and Moon around the Sun. So unless there is a good physical reason for the much smaller gravitational influences of the planets to have a proportionately greater impact, they are unlikely to have a significant effect on the climate. One such possibility is that they can produce potentially important perturbations on the Earth's orbit. When the Earth is on one side of the Sun and all the other planets grouped in a tight arc on the other side, there is evidence that this configuration has reproducible effects in Chinese climatic records. Known as a synod, this particular alignment of the planets occurs every 179 years or so, although every five or six cycles it can be as short as 140 years. The fundamental rhythm of these synods is defined by the approximate alignment of Jupiter, Saturn, Uranus and Neptune. The movements of Mercury, Venus, Earth and Mars define the time of year when the synod occurs.

A study of Chinese records examined those occasions when the remaining planets were grouped within an arc subtending less than 90° at the Earth. After 1600 BC in China, when the synod occurred in the summer half of the year, the subsequent few decades tended to feature warm summers. Conversely, after winter synods there were more frequent cold winters.

Moreover, where the grouping was narrower the effects tended to be more pronounced. The physical explanation for these observations appears to be in the way in which the planetary configuration causes the Earth to speed up or slow down in its orbit. While the period of the orbit (365.24 days) remains unaltered, when the Earth is travelling towards the grouping of giant planets it speeds up, and when travelling away from them it slows down. This means that it spends less time in the half of the orbit when it is closest to the conjunction of planets and more time on the far side of the Sun. So if the synod occurs in the summer half of the year this period will be lengthened slightly and vice versa. In the extreme example in 1665 when the planets subtended an angle of 45° to the Earth, it is estimated that the winter half of the year was increased almost two days with a corresponding shortening of the summer half of the year. This is potentially a significant physical shift and may, in part, explain why this synod marked the onset of the coldest period of the Little Ice Age in the northern hemisphere (see Section 4.9).

Another possibility is that the planetary motions influence the observed cyclic behaviour of sunspot numbers. The periods of the orbits of Jupiter, Uranus and Neptune roughly coincide with the 11-, 90- and 180-year cycles in the sunspot records and this has led to attempts to produce a planetary theory of sunspots. Of particular interest are the effects of the planets on the motion of the Sun around the centre of mass of the solar system. Calculations show that this complicated motion is dominated by the orbits of Jupiter and Saturn, and, in particular, the time taken for Jupiter to lap Saturn – 19.9 years. But over the last 1,200 years this period of the Sun's motion has varied between fifteen and twenty-six years. The other important cycle is 177.9 years which is a product of the near coincidence of fifteen Jupiter orbits and six Saturn orbits. The consequence of the Sun's motion is to affect its oblateness, diameter and rate of spin, all of which could influence the sunspot mechanism. So we cannot consider solar variability and tidal forces in isolation.

8.7   ORBITAL VARIATIONS

The Earth's orbit around the Sun is also influenced by the gravitational interactions with the Moon and other planets on much longer timescales. The resulting perturbations give rise to cyclical variations in orbital eccentricity, obliquity and precession with periods of 413, 100, 41, 19 and 23 kyr respectively (see Section 2.1.4). These variations are climatically important as they control the seasonal and latitudinal distribution of solar radiation. This theory of the climatic effects is usually attributed to the Yugoslav geophysicist Milutin Milankovitch, who transformed the earlier semi-quantitative work by James Croll into the mathematical framework of an astronomical theory of climate. This theory has been refined since the

1960s to provide an explanation of the observed waxing and waning of the ice ages over the last million years.

As noted in Section 2.1.4, the key to explaining how variations in the orbital parameters can trigger ice ages is the amount of solar radiation received at high latitudes during the summer. This is critical to the growth and decay of ice sheets. At 65° N this quantity has varied by more than 9% during the last 800,000 years. Fluctuations of this order are sufficient to trigger a significant climatic response. But, in explaining observed climatic change (see Section 4.4), there is a major snag. It is now accepted that latitudinal and seasonal variations in incident solar radiation due to the precession of the equinoxes (the 19- and 23-kyr periodicities) and the variations in the tilt of the Earth's axis (the 41-kyr periodicity) are sufficient to drive the significant climatic changes observed on these timescales. By comparison, the 100-kyr eccentricity periodicity is the weakest of the orbital effects. This causes considerable difficulties as the 100-kyr ice-age cycle is the strongest feature in the climatic record in the last 800,000 years. So it is necessary to consider separately the explanations of the observed 19-, 23- and 41-kyr cycles and the dominant 100-kyr cycle.

A direct approach to possible models of the ice ages was adopted by John Imbrie and John Z. Imbrie of Brown University, Rhode Island in 1980. Instead of using numerical models to test the astronomical theory, they used the geological record as the yardstick against which to judge the performance of various physical models. These models, which have become more sophisticated over the years, fall into two broad categories. The first group adopted an equilibrium approach to the changes in solar radiation. This involved calculating the climatic conditions that should occur for various combinations of orbital parameters. This produced reasonably realistic changes in temperature patterns, but was unable to reflect the inertia of the climatic system, notably the characteristic timescales of the growth and decay of the ice sheets, which appear to be of the same order of magnitude as those of orbital forcing.

The alternative approach is to use a differential model in which the rate of change of the climate is a function of both the orbital forcing and the current state of the climate. Not only is this a more realistic representation of climatic behaviour but it also contains a non-linear relationship between the input and the output which has important physical consequences. But it remains controversial as there is no agreement as to which climatic factors should be given particular emphasis and what values should be attached to them. For this reason the simple empirical approach used in the model developed by the Imbries will be considered first, as it provides valuable insights into the processes involved.

The Imbrie model considered only the link between the orbital forcing function and the land ice volume. This approach reflected the fact that the change in ice volume as recorded in the oxygen isotope records from deep-sea cores is the most accurately defined climatic parameter over the last million years or so, and also that the cryosphere is the part of the climatic system whose characteristic timescales of response closely match the dominant 100-kyr period of the orbital forcing. Indeed the common feature to emerge from all the modelling work in recent years is that only when the northern ice sheets exceeded a critical size did the 100-kyr cycle take over in the climate equation. This is seen as the key to explaining why before around a 800,000 years ago this cycle did not feature strongly in the climate records (see Fig. 6.15). The earlier interval of the Cenozoic ice ages, from 2.4 to 0.8 Myr ago, was almost completely dominated by the 41-kyr tilt cycle. Prior to this the 19- and 23-kyr cycles were more important. The advent of the 41-kyr cycle around 2.4 Myr ago is linked with the start of major northern hemisphere glaciation. So the lengthening period of the major climatic variations during the last five million years or so appears to be due to the increasing size of the ice sheets.

If the size of the northern ice sheets is critical, the important factor is the sensitivity of the ice volume to the time constants for the growth and decay of the ice sheets and the lag between the changes in the solar radiation falling in summer at high latitudes of the northern hemisphere. Once the Imbrie model had achieved a reasonable representation of this long-term behaviour, the more rapid response of the rest of the global climate to the other orbital parameters, which have a linear impact, could be added in to build up a better picture of the progression of the ice ages.

In spite of developments in recent years, the Imbrie model remains a good start to considering the physical processes which may be at work in the 100-kyr periodicity. In this model the parameters were tuned to achieve the best fit between the calculated ice-volume changes and the oxygen-isotope record. The most important features of the model include, first, the orbital forcing is fixed by the changes in the tilt of the Earth's axis and the precession of the perihelion of the orbit. The changes in eccentricity (i.e. the 100-kyr and 413-kyr periodicities) do not exert a significant influence on the seasonal and latitudinal variations of the radiation input. So the orbital forcing used in the model contained only the 19-, 23- and 41-kyr cycles, in spite of the fact that the ice-volume curve over the last 800,000 years is dominated by the 100-kyr cycle. The second essential feature is that the time constants of growth and decay of the ice sheets are markedly different. This reflects the evidence that during the last 800,000 years or so the ice sheets built up slowly, but collapsed dramatically at the end of each ice age (see Fig. 4.12).

The best results were obtained with a set of parameters that included time constants of growth and decay of the ice sheets of 42.5 and 16 kyr respectively, and a lag of 2 kyr between the orbital forcing and the response of the climatic state. This produced a good fit over the last 150 kyr (Fig. 8.7), although prior to this the match was less good. More importantly, the calculated changes in ice volume included 100- and 413-kyr periodicities, although the relative strengths were wrong, with the former being too

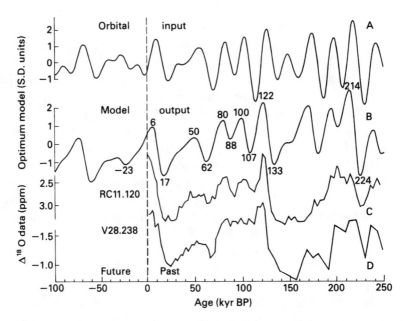

Figure 8.7 The combined orbital effects shown in Fig. 2.6 can be used as the input **(A)** to a model whose output **(B)** shows a marked similarity to the oxygen isotope variations observed in deep-sea cores from the southern Indian Ocean **(C)** and the Pacific Ocean **(D)** (Burroughs, 1994, Fig. 7.2).

weak and the latter too strong. But the fact that these essential periodicities were present at all highlights an important aspect of non-linear models. This is that the output spectrum has major features that are absent in the input forcing function. This is an example of the phenomena described in Section 8.1 where a simple non-linear system can generate sum and difference frequencies. The important feature here is that the choice of the time constants of ice growth and decay plus the non-linearity of the model combine to produce the required longer periodicities. The difference between the 19- and 23-kyr will produce a 110-kyr periodicity and that between the 23-kyr and 41-kyr will produce a 52-kyr periodicity, and the difference between these two is a 100-kyr cycle. Tuning the model and highlighting a given frequency is both rewarding and also underlines the limitations of the approach adopted. The dependence on empirically derived time constants, which have only the broadest links with the physical behaviour of the ice sheets, is a major limitation.

The process driving the 100-kyr cycle remains the key challenge in improving ice-age models. An overall assessment produced in 1993 of the many models developed to address this issue confirmed that the non-linear response of the global climate to the northern ice sheets exceeding a critical value was the explanation of the 100-kyr cycle. All that was required was that the build-up of the ice sheets introduced a lag of some 15 kyr into the system for the climate to cease to be dominated by the 23- and 41-kyr cycles and switch into the 100-kyr mode. Components of the atmosphere–ocean–ice system could be responsible for generating this response. The eccentricity cycle need play no part in these changes: they could be nothing more than a natural response of the climate to the variations in the size of the ice sheets.

An interesting refinement to this type of thinking has recently been proposed by Didier Paillard, at CEA/DSM in France. He has produced models which include the possibility of the climate system switching between three distinct climate regimes (e.g. interglacial, mild glacial and full glacial). These different states in the climate system could well be the product of different modes of global ocean thermohaline circulation (see Section 8.3). The switch between these regimes is controlled by a combination of changes in insolation and/or ice sheet volume. By defining the conditions for the transition between the three regimes the model is capable of reproducing with remarkable accuracy the changes in ice sheet volume over the last million years. So, yet again, the capacity of the global climate to move between markedly different, but relatively stable states appears to be an essential feature of explaining bigger climate changes.

Going beyond the interactions between the atmosphere, oceans and ice sheets, and the Earth's orbital parameters, other factors have been invoked to explain why the climate has become more susceptible to glaciation in recent geological history. In particular, it has been argued that the rapid tectonic uplift of the Himalayas and parts of western North America during the past few million years has increased the sensitivity of the global climate to orbital forcing. The link between the meandering of the jet stream, these mountainous areas (see Section 3.4) and the regions where the northern ice sheets developed may be the key to the current pattern of ice ages.

8.8   CHANGES IN ATMOSPHERIC COMPOSITION

By comparison with the complexities of possible consequences of low levels of cyclic perturbations on the climate, the long-term change in the amount of radiatively active gases in the atmosphere (see Section 2.1.3) looks like a relatively open and shut case. Clearly, fluctuations in these atmospheric constituents are bound to have exerted some influence on the climates of the past. The much higher levels of CO_2 in the atmosphere during, say, the Cretaceous (see Section 4.2) must have been an important factor in the maintenance of the benign climate of that period. The more difficult question is just how important a factor it was when compared to the issues of continental distribution and ocean circulation and what part the long-term decline in CO_2 over the last fifty million years or so has played in the general cooling of the climate. The broad conclusion of various computer-model studies (see Chapter 9) is that the location of the continents and the oceans are not a sufficient explanation, and that the long-term cooling over this period may be associated with a gradual decrease in atmospheric CO_2.

Of more immediate interest in terms of the current climatic debate (next section) is how the level of greenhouse gases fluctuated during the last ice age. There is the close parallel between both CO_2 and CH_4 levels observed in bubbles trapped in the Antarctic ice-core and temperature levels (see Fig. 6.14). The correlation between CO_2 levels and temperature over the last 220 ky is 0.81 (the figure for CH_4 is 0.76). Both these figures are statistically highly significant and are clear evidence that variations in the greenhouse gases played a direct part in the observed change in the climate.

These examples show that there is no doubt that the changing level of greenhouse gases in the atmosphere has been an important part of climate change during the Earth's history. The physical arguments supporting this statement will be explored in more detail in Section 10.2. Here the central question is how changes in the level of greenhouse gases have been linked with other causes of climate change. Because the levels of CO_2 and CH_4 are the product of activity in the biosphere which, in turn, is linked to prevailing climatic conditions, interpretation of the role of past variations in greenhouse gas levels in defining the climate have to account for how these feedback mechanisms operate. So, for instance, how the changes in the levels of these gases during the last few hundred thousand years have combined with orbital factors to produce periodic ice ages is a more complicated issue. Answers to such questions are, however, particularly important when using palaeoclimatic evidence in support for the predictions of how the build-up of greenhouse gases in the atmosphere due to the combustion of fossil fuels will lead to global warming (see Section 10.2).

In the case of ice-core data, when the changes in temperature are examined in closer detail, it becomes clear that there are subtle differences between the changes in the temperature and the rise and fall in the levels of CO_2 and CH_4. This suggests that other climatic factors were the primary cause of the shift in the climate, and the changes in greenhouse gases may have been no more than a secondary factor. Most notable is the fact that, while CO_2 levels rose in line with rising temperature, they did not follow it down into the glacial conditions. Instead they remained at relatively elevated levels for several thousand years at the end of the last interglacial, and then descended in two stages to their lowest levels. One explanation for the intermediate level over the period from around 115 to 75 kyBP was that the oceans continued to maintain a high level of biological activity until around the end of Stage 4 (see Table 4.2). So, while the greenhouse gases acted as part of a positive feedback mechanism, which reinforced the changes during the last ice age, these events may not be a useful guide to what may happen as a result of human emissions of these gases: as with every aspect of climate change, life is never as simple as it at first appears.

8.9 HUMAN ACTIVITIES

How human activities may be altering the global climate is considered in detail in Chapters 9 and 10. But, in terms of a general review of the causes of climatic change, a brief analysis of how various of these activities could have already had an impact needs to be covered here.

At the local level there is no doubt human activity alters the climate. As discussed in Section 6.1, correcting for the effects of urbanisation is a major factor in obtaining a reliable measure of temperature trends. Apart from the warming effect of cities, they also reduce wind speeds, reduce visibility by the formation of particulates and photochemical smogs and, in certain circumstances, increase the chances of heavy precipitation. These effects are, however, restricted to a very small part of the Earth's surface and so far their global consequences are negligible. But they do provide insight into how human activities can produce a complex web of climatological impacts.

On a global scale the atmospheric consequences of human activities occur in a number of areas. The most closely studied are the emissions of various radiatively active gases which are leading to an increase in the greenhouse effect (see Section 2.1.3). Pride of place goes to CO_2. Its atmospheric concentration has been monitored closely since 1958 at Mauna Loa in Hawaii (Fig. 8.8) and, for nearly as long, at a variety of other sites, including the South Pole. When combined with ice-core data the conclusion is that CO_2 levels have risen from about 280 parts per million by volume (ppmv) in the pre-industrial period (defined as the average of several centuries before 1750) to around 360 ppmv in the late 1990s – the precise figure varies around the globe and with the time of year, as the growing season in the northern hemisphere has a dominant influence on the annual cycle (Fig. 8.9). While this rise appears to have grown inexorably, closer examination of the Mauna Loa data shows that the rate has fluctuated dramatically with rapid growth rates in the late 1980s and a marked slow down in the early

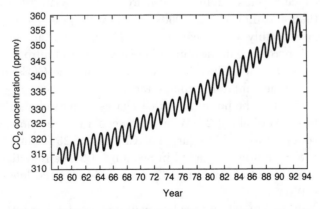

Figure 8.8 CO_2 concentrations measured at Mauna Loa, Hawaii, since 1958 showing trends and seasonal cycle (IPCC 1994, Fig. 1.3).

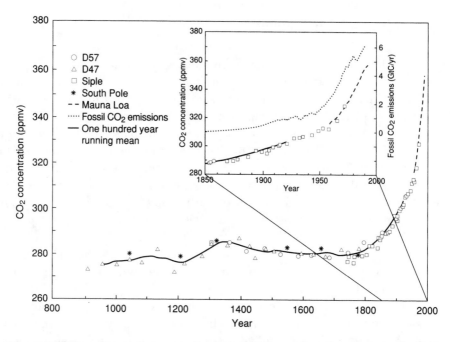

Figure 8.9 CO_2 concentrations over the last 1,000 years from ice-core records (D47, D57, Siple and South Pole – all in Antarctica) and (since 1958) Mauna Loa, Hawaii, measurement site. The smooth curve is based on a 100-year running mean. The rapid increase in CO_2 concentration since the onset of industrialisation is evident and has followed closely the increase in CO_2 emissions from fossil fuels (see inset of period from 1850 onwards) (IPCC, 1995, Fig. 1(a)).

1990s (Fig. 8.10). These fluctuations have not been explained but suggest complicated feedback mechanisms between short-term climatic variations (e.g. the ENSO) and the up-take of carbon in the biosphere.

Analysis of CH_4 concentrations show a similar pattern, having risen from a pre-industrial level of around 700 parts per billion by volume (ppbv) to about 1730 ppbv by the end of the 1990s. Again the rate of growth has fluctuated in recent years having slowed from 15 ppbv in 1980 to about 10 ppbv in 1990 and then in the 1990s has slowed appreciably, although fluctuating considerably from year to year. Other important greenhouse gases which are increasing in the atmosphere are oxides of nitrogen, notably nitrous oxide (N_2O), and halocarbons (these include the CFCs and other chlorine and bromine containing compounds).

The overall effect of the build-up of these gases is to produce a radiative forcing (see Box 2.1) of about 2.5 W m^{-2} between 1850 and now. CO_2 has contributed some 60% of this figure and CH_4 about 25%, and N_2O and halocarbons the remainder. As noted in Section 2.1.3, the radiative forcing owing to the equivalent of doubling CO_2 from pre-industrial levels is estimated to be 4 W m^{-2}.

Other areas of human activity are less well understood. For instance, the formation of atmospheric particulates and changes in cloudiness is a murky

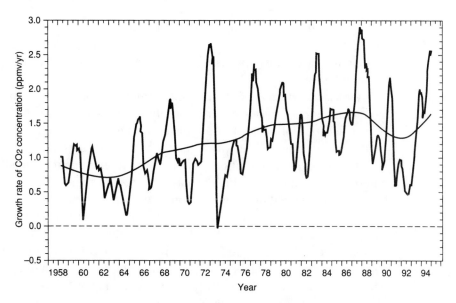

Figure 8.10 Growth rate of CO_2 concentrations since 1958 in ppmv per year at Mauna Loa, Hawaii, together with a smoothed curve to show variations on timescales longer than around ten years (IPCC, 1995, Fig. 2.2).

business. Their impact on the radiative balance of both incoming solar radiation and outgoing terrestrial radiation depends on their size, shape and absorptivity. This means whether particulates absorb, reflect or scatter radiation is a complex function of their properties.

A good example of these effects is the formation of sulphate particulates as a consequence from the combustion of sulphur-containing compounds in fossil fuels. The resulting emissions of sulphur dioxide are converted to sulphuric acid in the atmosphere which then forms sulphate aerosols that can both absorb and reflect sunlight in their own right. Their net effect will depend on the underlying surface. Over low albedo surfaces (e.g. the oceans and forests) they will probably increase the amount of sunlight reflected to space. Over high albedo surfaces (e.g. snow and deserts) they may be net absorbers. They will also have, as yet, unquantified indirect effect of modifying the properties of clouds. It is proposed that these effects have delayed the global warming anticipated with the build-up of greenhouse gases in the atmosphere during the twentieth century (see Section 10.2).

Another form of particulate emissions is the direct result of altering the surface of the land. Where agriculture has removed forests and exposed the soil there is an increased chance of dust being swept up into the atmosphere, especially during times of drought. It is estimated that the overall impact of altering the Earth's surface and injecting extra dust into the atmosphere has had a significant cooling effect. It could even be sufficient to cancel out much of the impact of the build-up of greenhouse gases in the last century. Conversely, increased atmospheric dustiness at the end of the last ice age caused by strengthening global circulation could have accelerated the warming process as the dust would have absorbed more sunlight than the ice sheets beneath it. So, as with so many other aspects of the climate, the response is non-linear and depends on the state of other climatic factors.

The associated issue of whether these processes lead to the more permanent formation of desert areas (*desertification*) is the subject of an unresolved debate. Following the prolonged drought in the Sahel in the early 1970s (see Section 5.8), it was argued that the increased albedo associated with the removal of vegetation would lead to a permanent expansion of the desert. To the extent that nomadic pastoralists with their flocks were responsible for the removal of the vegetation, it was argued that the desertification at the time was the result of human activity. Subsequent satellite observations showed that the extent of the Sahara Desert was closely linked to fluctuations in rainfall, and vegetation cover rapidly regenerated in wetter years. Now, it is recognised that desertification is a complex process in which climatic change is probably the dominant factor and human activities play a relatively minor part (see Further Reading).

The related question of other consequences of human activities modifying the land surface has concentrated on the impact of deforestation. The subject of the destruction of tropical forests has attracted most attention. The arguments for this leading to appreciable global climate change are, however, not compelling as the impact of the reduction in albedo may be compensated by a decline in cloudiness as the land dries out. By comparison, the impact of removing northern forests may be more substantial. The reason is all to do with snow. Snow-covered forest absorbs about half the sunlight falling on it, whereas snowy, open ground absorbs only about a third. So, the effect of widespread deforestation would be to produce a dramatic cooling across the snowy higher latitudes of the northern hemisphere, especially at the end of winter and in early spring. Both computer climate models and practical developments in numerical weather prediction (see Section 9.1) suggest that this could lead to cooling by as much as 10 °C where large areas of forest are removed. This sensitivity may explain the warmth of the Holocene maximum (see Section 4.6) when the northern forests extended farther north and their presence would have accelerated the melting of snow and increased spring warmth at high latitudes.

8.10   CATASTROPHES AND THE 'NUCLEAR WINTER'

Few subjects in climatic change excite greater passions than the possibility of the Earth being hit by an extraterrestrial object. Such a catastrophe, whether as the result of a collision with a large meteorite, asteroid or fragments of a comet, is regarded by many as being in the realms of science fiction. Part of the reason for this reaction is that it is sometimes seen as undermining the principle of uniformitarianism (see Section 4.1) which was originally conceived to deal with earlier explanations of the geological record. These had placed great emphasis on catastrophes being the explanation for many features of the landscape. But, as presented here, the occasional impact from an asteroid is not seen as undermining the principle of uniformitarianism in any real sense.

The simple fact of the matter is that there can be little doubt that an object greater than, say, a kilometre across would have a massive temporary impact on the climate. Equally well, there can be no doubt that the Earth has been struck by objects of this size and greater in the past. What is at issue is whether these events can explain past changes in the climate and what is the probability of a significant event occurring in the foreseeable future.

The most recent example of the Earth being hit by a sizeable object was on 30 June 1908 in Siberia. This was either a meteorite or a fragment of a comet some 50 m across. It burnt out at an altitude of around 10 km over the Tunguska River, about 1,000 km north-north-west of Lake Baikal, with an explosive force equivalent to a 10–20-megaton bomb. It devastated 2,000 km^2 of forest and generated a seismic shock of about five on the Richter scale. While the climatic impact of this event was small, it clearly demonstrates the potential of collision of the Earth with objects in space. Equally impressive is the impact of the fragments of the comet Shoemaker–Levy 9 with Jupiter in July 1994 providing a graphic illustration of the consequences of such an event. The biggest fragment was about 3 to 5 km across, ejected superheated gas to a height of 1,300 km and created a circular cloud larger than the entire Earth. Although there are a variety of physical arguments about the relative impact of different types of objects (e.g. meteorites with a variety of compositions, or cometary fragments – the generic term *bolide* is used to describe these objects), the message is simple – any body of a kilometre or more across would do immense damage and have a profound and immediate impact on the climate.

As noted in Section 5.3, the possibility of bolide collisions being the cause of some or all of the major mass extinctions is a matter of vigorous controversy. The most comprehensively analysed event relates to the end of the Cretaceous, and the passing of the Age of the Dinosaurs. These arguments about the scale, pace and nature of the extinction that occurred at around 65 Ma will continue to rage for a long time. But, what we need to consider is the climatic consequences of the proposed bolide collision. The accepted view is that a crater on the Yucatan peninsula in Mexico is the site of impact. A body some 10 km across hit this part of the world making a crater between 200 and 300 km across. The immediate consequences of this impact was cataclysmic. The combination of massive earthquakes, huge tidal waves, widespread forest fires and vast amounts of dust and debris hurled high into the atmosphere would have done untold damage. What is more pertinent is whether this produced lasting climatic effects which could have precipitated the extinction of the dinosaurs?

The answer may lie in work that was done for entirely different reasons. During the late 1970s a lot of studies were conducted into the climatic impact of nuclear warfare. These considered the range of environmental consequences of a large-scale nuclear exchange between the major powers involving the detonation of the equivalent of some 6,000 megatons of nuclear devices. They estimated the direct consequences of nuclear explosions, the subsequent fires and smoke generation and the impact of this smoke on the global climate. This combination was predicted to produce a sudden and dramatic cooling which was widely known as the *nuclear winter*. Many features of this analysis can be translated into the catastrophic consequences of a bolide impact. This is because there is some comparability between the events. A kilometre-diameter bolide would have an impact energy of 100,000 megatons, while a 10-km object, comparable to the one that hit 65 Ma, would be equivalent to 100 million megatons. Although the predicted impact of a large bolide is much greater, its impact is concentrated in one spot whereas a nuclear conflict would consist of many much smaller explosions spread over a much wider area. But, such an impact would eject huge amounts of material high above the Earth. This would then rain down like molten meteorites over a vast area, setting fire to forests and grasslands. So the analogy between a nuclear winter and these prehistoric catastrophes may be a useful approximation. Indeed there is evidence of global fires in the form of large amounts of soot present in sediment marking the Cretaceous–Tertiary extinction boundary.

There is another feature of this boundary which fits in well with nuclear winter hypothesis. This is the sudden drop in pollen from flowering plants and a jump in fern spores. This is known as the *fern spike*. In a few millimetres of the sedimentary sequence, the fern spore content goes from 25% to 99%. This change is reminiscent of sudden shifts of vegetation often seen after a forest fire today – for a few years, a lush forest is replaced by an opportunistic flora dominated by ferns.

The broad conclusion is that a nuclear winter would be a dramatic, but short-term event. If it occurred in the summer half of the year in the northern hemisphere it would cause a drop in continental land-surface temperature of between 20 and 40 °C within a few days. Thereafter, the smoke clouds would stabilise and might be held aloft for as much as a year, thereby maintaining winter-like conditions over the northern continents for at least one growing season and possibly longer. This more general cooling would have a global impact, causing frosts in regions that normally never experienced them. So the overall consequences on all forms of land-based life would be catastrophic, with only the most adaptable surviving.

In terms of climate change such an event can be regarded as both extraordinarily dramatic but also exceedingly transient. Unless other parts of the climate system experienced some more profound shift, it would soon return to normal as the atmosphere cleared of smoke and debris. One such change might be a sustained coincident period of volcanism. There have been a number of occasions in geological history when such events have occurred. Lasting hundreds of thousands of years, they could have a huge impact on the atmosphere and maintain an atmospheric dust veil (see Section 8.4) for long enough to precipitate a longer cold episode with the expansion of the polar ice-sheets. There is evidence of massive volcanism at around the end of the Cretaceous, which may have combined with a bolide impact to change the climate. Indeed, there is no reason why either the bolide impact or volcanism alone has to be the sole explanation of the extinction, they could both have contributed to the death of the dinosaurs.

On the more general question of the incidence of impacts and their distribution over time the arguments rage equally fiercely. Two issues matter. The first is to establish just how often impacts of different sizes can occur. The second is to decide whether there are good reasons for there being any regularity in the spacing of the biggest impacts. The frequency of impacts is no easy matter as most objects burn up in the atmosphere and leave no detectable evidence of their arrival. Even the Tunguska impact left no clearly identifiable trace of its make-up, after its explosive arrival. So usually only the biggest objects produce craters. The notable exception is iron meteorites which often reach the surface and if sizeable produce distinctive craters (Fig. 8.11). Furthermore, identifying major impacts in geological history is difficult as many features have been covered up by sediment, eroded away or lie beneath the oceans. For this reason, the estimates of the current rate of impacts of different sizes is based on the analysis of new small craters on the Moon (Fig. 8.12), which show up because of the bright rays of ejecta around them. This suggests a Tunguska event might happen every few hundred years, while a climatically significant event of, say, about 1,000 megatons could occur every 10,000 years or so.

The question of whether collisions will be random or exhibit some regularity depends on how debris is distributed in the vicinity of the solar system, in general, and in Earth-crossing orbits in particular. Asteroids and meteorites can probably be regarded as randomly distributed, although the variations in the frequency of micrometeorite impacts (*shooting stars*) throughout the year suggests that there are regions of higher density in the vicinity of the Earth's orbit. Cometary debris is more likely to be located in distinctive patterns related to the orbits of massive comets which have broken up in the past. This whole issue of where these orbits are, in respect of the Earth, and how frequently our planet will run into this type of debris is the subject of controversy. Suffice it to say, if there are regions in space where such objects are more common, when the Earth moves through these regions the risk of impact will rise.

Leaving aside the questions about the more distant prospect of mass extinctions whose frequency is measured in tens of millions of years, there is the question of whether any events in recorded history could be

Figure 8.11 The meteorite crater in Arizona which was formed some 50,000 years ago. The diameter of the crater is 1.2 km (*Cambridge Encyclopaedia of Earth Sciences*, Fig. 2.13).

attributed to smaller impacts. Perhaps the best candidate for such an event is the 'mystery cloud' of AD 536. It is recorded by chroniclers from Rome to China that the Sun dimmed dramatically for up to eighteen months and widespread crop failures occurred. These contemporary observations are supported by clear evidence of climatic deterioration in tree-ring widths (Fig. 8.13). This event shows up much more strikingly in tree-ring data than the impact of the volcano Tambora in 1815 (see Section 8.4).

The most commonly assumed explanation is that this event was the result of a major volcanic eruption, possibly Rabaul in New Guinea, which created a global dust veil in the upper atmosphere that led to a cooling of the Earth's surface (see Section 8.4). But the evidence of the eruption in Greenland ice cores (see Section 6.4.2) is, at best, ambiguous: an event at around this time does not match up to either Tambora, or an unidentified eruption in AD 1259. In the absence of confirmation of an exceptional eruption around AD 536, it can be argued that an alternative explanation is needed and a collision with an extraterrestrial object is a plausible option. This debate will only be resolved by new measurements in, say, ice cores which point unequivocally in one direction or other. In the meantime, all that can be said is that there is

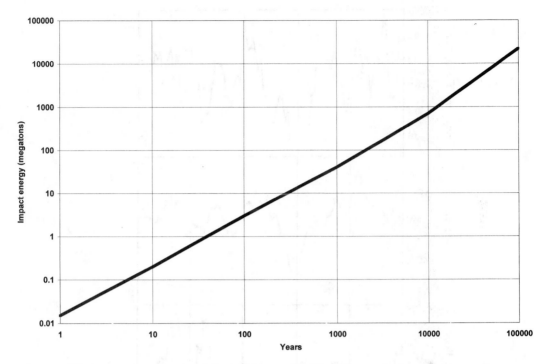

Figure 8.12 An estimate of the incidence of bolide impacts of given energies on the Earth (data from Shoemaker, 1983).

no doubt there was a cataclysmic event in AD 536, which had far-reaching historical consequences (see Section 5.7), and it is possible that it was caused by an extraterrestrial object.

8.11   SUMMARY

This overview of the principal causes of climate variability and climate change provides a flavour of the challenges facing anyone wishing either to explain past events or to predict future developments. The combination of the wide variety of physical processes that could produce fluctuations and the fact that many of these are difficult to distinguish from one another means it is not easy to decide which effects matter most. But, we must take a view on the past and make plans for the future. This requires decisions as it is not feasible to try to include all the possible causes of change in any analysis. So the message is to draw out the leading factors from the areas covered in this review.

Two priorities emerge from this review. First the most pressing problem is to get a better fix on the internal variability of the climate. Finding out more about how the oceans and the atmosphere interact on timescales of decades and centuries is central to deciding to what extent changes during

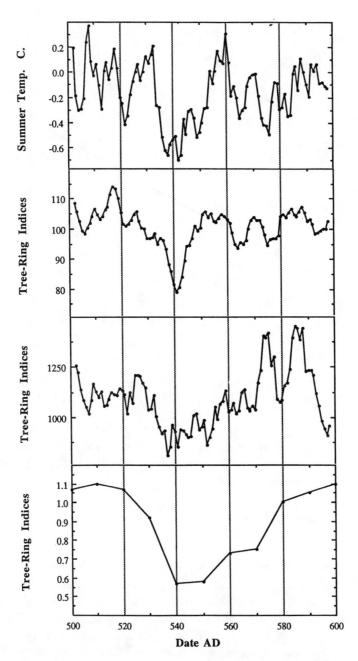

Figure 8.13 Tree-ring observations from various parts of the northern hemisphere showing a consistent picture of markedly cold summers around AD 540. The curves from top to bottom are the five-year running means of Fennoscandian summer temperatures, European oak width indices, and bristlecone width indices from the White Mountains in California, together with decadal foxtail pine width indices from the Sierra Nevada Mountains in California (Baillie, 1995, Fig. 6.4).

the twentieth century are the product of climatic autovariance as opposed to human activities. In addition, providing an explanation as to why the climate of the last 10,000 years has been relatively stable and whether this stability could be disturbed by natural causes or human activities is equally important.

The second, and related priority is to establish a balanced and more thorough understanding of how human activities can affect the climate. This is not just a matter of increasing levels of greenhouse gases, but also the direct and indirect effect of particulates and the changes in land use including desertification and deforestation. At the same time more research must be conducted into the impact of other factors (e.g. volcanoes, solar activity and tidal effects). But these are subordinate issues, although a growing knowledge of how these factors alter the climate will assist in the primary objectives of understanding the nature of autovariance and the impact of human activities.

Progress on these priority areas will depend on both improved measurements of the type described in Chapter 6, and better computer models. So now we must move on to the question of modelling the climate.

QUESTIONS

1 Given the non-linear properties of the climate, explain why we need to consider the dominant nature of the annual cycle in our analysis of weather cycles. What periodicities could be generated between this cycle and the QBO (see Fig. 3.9), a three- to five-year ENSO signal, and the eleven- and twenty-two-year solar cycles? What do your calculations tell you about the value of putative cycles in forecasting the future climate?

2 What are the differences between high clouds and volcanic dust injected into the upper atmosphere which mean that high clouds warm the climate while volcanic dust has a cooling effect?

3 What are the arguments for and against a policy which proposes that many of the climatic consequences of human activities may compensate each other and, hence, we do not need to take drastic action to prevent climate change?

FURTHER READING

A complete reference list is available at the end of the book but the following is a selection of the best books or articles to follow up particular topics within this chapter. Full details of each reference are to be found in the Bibliography.

Bigg (1996). Particularly valuable in providing analysis of the growing understanding of how the atmosphere and oceans combine to govern many aspects of the climate.

Bryant (1997). Well worth reading as it adopts a somewhat different position to, say, the IPCC on the relative importance of various causes of climate change and our ability to estimate their impact.

Imbrie and Imbrie (1979). An accessible presentation of the research into the causes of the ice ages, and how new thinking developed during the 1960s and 1970s.

Imbrie et al. (1992) & (1993). These papers review the vast amount of work that has been done on modelling the response of the Earth's climate to orbital forcing and provide a good assessment of our current state of knowledge.

Open University Oceanography Series (1989). The ocean circulation volume provides particularly clear introduction to the basic aspects of ocean dynamics.

Pecker and Runcorn (1990). A collection of papers which provides a particularly comprehensive review of the nature and origin of solar variability and a set of interesting observations about how this behaviour may be linked to climatic change.

Philander (1990). A thorough and penetrating survey into the nature and causes of large-scale climatic fluctuations in the tropical Pacific and their influences on global climate.

Shoemaker (1983). The definitive analysis of the likely incidence of significant bolide impacts with the Earth.

Thomas and Middleton (1994). A stimulating and controversial analysis of the causes of desertification which is required reading for anyone who wishes to understand the complexity of this difficult subject.

CHAPTER NINE

MODELLING THE CLIMATE

Vaulting ambition, which o'erleaps itself
Shakespeare (*Macbeth*)

Attempts to produce models which accurately reflect the complexity of the Earth's climate is an immense challenge. Even when focusing on the priority areas of natural variability and the impact of human activities, the task is daunting. Extending it to cover all the potential causes of climate change may prove overwhelming. So the approach must be to develop systems which inspire confidence in our ability to address the priority areas, and which can then explore the sensitivity of the climate to other factors. The only way to do this is to create detailed computer models of the global climate, to evaluate the various hypotheses. The first stage is to establish whether these models are capable of producing a realistic representation of the climate and can respond to the most obvious quantifiable perturbations. The next stage is to check the relative importance of other possible causes of climate change, both to ensure we have correctly identified the priority areas and also to gain a better understanding of past events and decide how much confidence to attach to predictions of future changes.

Given these broad aims of climate modelling, the objective of this chapter is to describe the essential features of computer models of the climate and then to assess their performance in terms of reproducing known features of the climate. After this, the ability of the models to handle the natural variability of the climate and the possible impact of human activities will be considered. In the light of this analysis, the challenges facing modellers will be discussed in terms of whether reliable predictions of climate change are a realistic proposition within the foreseeable future.

9.1   GLOBAL CIRCULATION MODELS

In considering climate models, the obvious starting point is the numerical weather predictions which provide our daily forecasts. Because we are so familiar with these products of computer models, it is easy to relate to their performance. Weather forecasting is a problem in mathematical physics. The computer models used for forecasting contain a very large and complex array of equations based on the physical and dynamical laws which govern the birth, growth, decay and movement of weather systems. They incorporate the principles of conservation of momentum, mass, energy and water in all its phases; the Newtonian equations of motion applied to air masses, the laws of thermodynamics and radiation for incoming solar energy and outgoing heat radiation, and equations of state of atmospheric gases. Parameters which are specified in advance include the sizes, rotation, geography and topography of the Earth, the incoming solar radiation and its diurnal and seasonal variations, the radiative and heat conductive properties of the land surface according to the nature of the soil, vegetation and snow and ice cover, and the surface temperature of the oceans.

The physical state of the atmosphere is updated continually drawing on observations from around the world using surface land stations, ships, buoys and, in the upper atmosphere using instruments on aircraft, balloons and satellites. Just assimilating this data involves four trillion calculations per forecast. The model atmosphere is divided into thirty or more layers between the ground and 30 km and, in the most advanced models, each level is divided up into a network of points about 50 km apart – some four million points in all. Each of these points is assigned a new value of temperature, pressure, wind and humidity with each run of the model and the governing differential equations are integrated in fifteen minute steps at each point to provide forecast values up to ten days ahead. Each new set of forecasts involves at least twenty trillion calculations. The output is hundreds of forecast charts of pressure, temperature, wind, humidity, vertical motion and rainfall which are used to provide a variety of forecasts.

So the scale of weather forecasting provides a good indication of the resources that can be brought to bear on the issues of climatic change. But, as we all know, weather forecasts have obvious limitations. While they have shown significant improvements in the last twenty years or so, they only provide useful forecasts out to around six days ahead. This slow progress highlights the unpredictable nature of the atmosphere. Of particular interest in terms of the longer term behaviour of the climate is the handling of regimes (see Sections 3.2, 3.6 and 7.2). Standard weather forecasts do not predict sudden switches between stable circulation patterns well. At best they get some warning by using statistical methods to check whether or not the atmosphere is in an unpredictable mood. This is done by running the models with slightly different starting conditions and seeing whether the forecasts stick together or diverge rapidly. This *ensemble* approach provides a useful indication of what modellers are up against when they seek to analyse the response of the global climate to various perturbations and to predict the course it will follow in the future.

To start with, the general circulation models (GCMs) used by climate modellers cannot represent the global climate in the same detail as the numerical weather predictions because they must run their simulations for decades and even centuries ahead in order to consider possible changes and not just ten days. Because of the high cost of running the most advanced computers and the simple limitations of time taken to complete the calculations, the models have to use a lower spatial resolution. Typically, the most advanced models have a horizontal resolution of 2.5° of latitude and 3.75° of longitude, but retain the detailed vertical resolution, having around twenty levels in the atmosphere. So while they use the same set of mathematics and physics as weather forecasting models, they cannot analyse atmospheric patterns to the same degree.

Climate models face a major additional challenge in handling the behaviour of the oceans. For weather forecasts a few days ahead the sea surface temperatures can be regarded as effectively constant and can use current values. But for climate modelling, the consequences of atmospheric–ocean interactions (see Section 8.2), are central to the whole business. The models have, therefore, to include realistic representations of the oceans and how they combine with the atmosphere in the process of climatic change. This poses a series of severe tests for modellers. The first of these is the fact that many of the important features of the oceans, notably currents and their associated eddies, are much smaller than the equivalent atmospheric systems. So, the models of the oceans should, ideally, have much higher resolution. The highest resolution research models of the oceans have sixty vertical levels and a horizontal resolution as fine as one sixth of a degree of latitude and longitude. Such detail cannot be incorporated into GCMs, which must calculate the rapid changes of the atmosphere as well as simulating longer term atmospheric–ocean dynamics, as it would take impossibly long times on even the fastest computers. This means, in practice, the oceans are considered at about the same resolution as the atmosphere and with around twenty vertical levels.

The compromise on resolution raises a fundamental problem in handling the physics of energy transfer between the atmosphere and the oceans where the two have markedly different temperatures. This boils down to how the models deal with fluxes of heat, momentum and freshwater between the oceans and the atmosphere. These will vary dramatically throughout the year and from place to place. For instance, the oceans will provide large amounts of heat to atmosphere in winter and absorb it in summer. These exchanges involve very large physical effects, and inevitably small errors build up over time, leading to the models drifting away from climatic normality. To prevent this happening some models make pragmatic adjustments, known as 'flux adjustments', to keep them on the climatic straight and narrow.

The performance of the best GCMs has been assessed in the 1995 IPCC Report. It considers the sixteen most advanced models which have been developed by the leading research groups around the world. The essential feature of this analysis is that in tackling the immense complexity of the global climate the modellers have adopted a variety of ingenious approaches to simplifying their computational task. This multiplicity of approaches provides insight into both which processes make a big difference to how the models perform and where the models have the greatest difficulty simulating the real world. Also the scatter in the results they obtain is a measure of their performance.

The simplest measure of their performance is the global figures they predict for temperature and precipitation. The figures for the eleven coupled atmospheric–ocean GCMs is given in Table 9.1. The striking feature of these results is that, while on average there is good agreement with observed values, the scatter between different models of nearly 6 °C is large compared with both past climate change and predicted future changes. Three of the four warmest models are from simulations without flux adjustment, with the flux-adjusted models generally producing a better simulation of surface air temperature over the oceans.

When the output of the models is examined in more detail the discrepancies become more striking. The latitudinal variation of simulated temperature from the actual values (Fig. 9.1). The biggest discrepancies are at high latitudes and over the continents. By contrast, differences over the ocean are rather small, with the greatest scatter being found over the Southern Ocean. The scale of these differences (see Fig. 9.1) indicates that some models have large climate drifts in this region.

The simulation of precipitation produces a comparable scatter. The broad features of peaks associated with tropical rainfall and the mid-latitude storm tracks appear in all the models. The variations are related to the intensity of rainfall and are largest in the tropics. These discrepancies are important as the precipitation rate is a measure of the intensity of the hydrological cycle and also influences the ocean thermohaline circulation (see Section 8.3). More generally the models with higher average temperature (see Table 9.1) have the larger precipitation rates, as might be expected, as this should increase the hydrological cycle.

The seasonal average mean sea-level pressure is represented rather well in coupled GCMs (Fig. 9.2). When it comes to simulation of snow and sea ice cover the models do, however, have considerable difficulties (see Tables 9.2 and 9.3). Some models have an excess of winter snow cover which persists into summer, while others, without flux adjustment, lose all their sea ice in the Arctic and almost all their sea ice in the Southern Hemisphere throughout the year. Clearly, the failure to simulate accurately this essential feature of the climate (see Section 3.4) is a major limitation.

TABLE 9.1 COUPLED MODEL SIMULATIONS OF GLOBAL AVERAGE TEMPERATURE AND PRECIPITATION

Institution	Surface Air Temperature (°C)		Precipitation (mm/day)	
	DJF	JJA	DJF	JJA
1. Bureau of Meteorological Research Centre (BMRC), Australia[a]	12.7	16.7	2.79	2.92
2. Canadian Centre for Climate (CCC)	12.0	15.7	2.72	2.86
3. Center for Ocean, Land and Atmosphere (COLA) USA[a]	12.6	15.5	2.64	2.67
4. Commonwealth Scientific and Industrial Research Organisation (CSIRO), Australia	12.1	15.3	2.73	2.83
5. Geophysical Fluid Dynamics Laboratory (GFDL), USA	9.6	14.0	2.39	2.50
6. Goddard Institute for Space Studies (GISS), USA[a]	13.0	15.6	3.14	3.13
7. Max Planck Institute for Meteorology (MPI), Germany (two models)	11.0 11.2	15.2 14.8	— 2.64	— 2.73
8. Meteorological Research Institute (MRI), Japan	13.4	17.4	2.89	3.03
9. National Center for Atmospheric Research (NCAR), USA[a]	15.5	19.6	3.78	3.74
10. UK Meteorological Office (UKMO), UK	12.0	15.0	3.02	3.09
Observed	12.4	15.9	2.74	2.90

[a] Models without flux adjustment.

The problems of flux adjustment also show up in the handling of ocean currents. There are substantial differences in how big a correction is made by the models, especially in the vicinity of ocean currents. The scale of this problem is best defined in terms of the models represented in Table 9.1. Their estimates of the poleward water transport in the North Atlantic by thermohaline circulation produce values covering the range 2 to 26 Sv $(1 \text{ Sv} = 10^6 \text{ m}^3 \text{ s}^{-1})$.

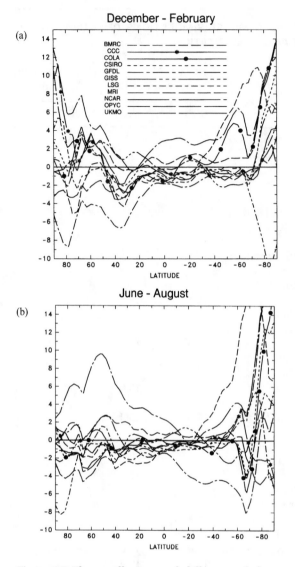

Figure 9.1 The zonally averaged difference of eleven coupled models' (see Table 9.1) surface air temperatures from climatological observations for **(a)** December–February and **(b)** June–August (IPCC, 1995, Fig. 5.1g and h).

Within this range only four of the ten models have circulation strengths which fall within the climatologically 'accepted' range of 13 to 18 Sv near 50° N. The fact that these four more successful models have both flux adjustment and a lengthy spin-up to an equilibrium state suggest these are necessary, but not sufficient, conditions for handling thermohaline circulation.

The ability of GCMs to handle clouds is another critical test. The importance of clouds on the Earth's radiation balance (see Section 3.3) means any inaccuracies in their simulation could have major implications for the reliability of the models. The most fundamental issue is whether climatic change

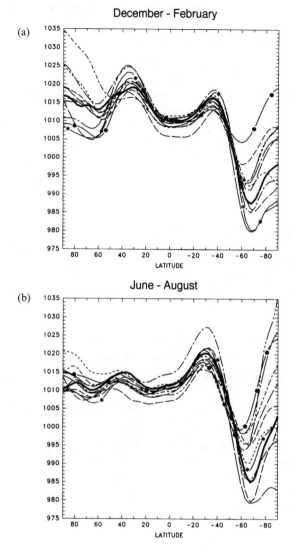

Figure 9.2 The zonally averaged mean sea level pressure from eleven coupled models (see Table 9.1) and that from climatological observations (solid line) for **(a)** December–February, and **(b)** June–August (IPCC, 1995, Fig. 5.3g and h).

will lead to coherent changes in cloudiness which reinforce any change (*a positive feedback mechanism*) or damp down the change (*a negative feedback mechanism*) (see Box 1.1). Because clouds exert such a strong influence on both incoming solar radiation and outgoing terrestrial radiation, the treatment of cloud type and distribution can have a major impact on whether the feedback is positive or negative. During the 1980s the models tended to predict that global warming would increase tropical cloudiness with a marked increase in the greenhouse effect in the tropics. The overall effect was that most models predicted a strong positive feedback so global warming would be

TABLE 9.2 COUPLED MODEL SIMULATIONS OF NORTHERN HEMISPHERE SNOW COVER (10^6 km^2)

	DJF (Winter)	JJA (Summer)
BMRC[a]	62.0	28.4
CCC	42.9	11.5
COLA[a]	53.4	12.1
CSIRO	37.5	11.6
GFDL	64.4	10.0
GISS (1)[a]	41.2	2.5
MPI (OPYC)	54.6	16.7
MRI	34.3	2.9
NCAR[a]	41.4	2.1
UKMO	35.3	5.1
Observed (Matson et al., 1986)	44.0	7.8

For details of institutions see Table 9.1.
[a] Models without flux adjustment.

TABLE 9.3 COUPLED MODEL SIMULATIONS OF SEA ICE COVER (10^6 km^2)

	Northern Hemisphere		Southern Hemisphere	
	DJF (Winter)	JJA (Summer)	JJA (Winter)	DJF (Summer)
BMRC[a]	18.9	16.7	<1.0	<1.0
CCC	9.7	7.1	12.2	7.5
COLA[a]	9.3	1.6	4.0	3.7
CSIRO	16.6	14.5	21.1	18.9
GFDL	16.0	12.7	24.7	16.0
GISS (1)[a]	11.0	8.3	12.4	6.4
MRI	19.1	11.5	11.6	3.1
NCAR[a]	13.6	<1.0	5.3	<1.0
UKMO	10.2	5.3	18.0	5.9
Observed	14.8	10.6	16.4	6.1

For details of institutions see Table 9.1.
[a] Models without flux adjustment.

further enhanced by increased tropical cloudiness. More recent models have produced far more ambiguous results, but this shift and the size of sensitivity of the new models to the *parameterization* of the clouds suggests that, without better figures, the simulation of the clouds is a major weakness in any prediction of climate change. In particular, the process of parameterization, which attempts to produce a reasonable measure of clouds by tuning various model parameters to produce the right mix of properties (e.g. the mixture of liquid and ice particles and the vertical and horizontal extent of clouds) will have to be replaced by better physical representations of their properties.

In terms of climatology the current models do only a modest job of portraying the latitudinal and seasonal distribution of global cloudiness (see Fig. 9.3). They systematically underestimate the cloudiness in low and middle latitudes in both winter and summer. In higher latitudes the models overestimate cloudiness, especially over Antarctica, even allowing for the uncertainty of satellite measurements in discriminating between clouds and underlying snow and ice in these parts of the world.

There is yet one more problem with modelling clouds. This is the question about whether global warming will alter the precipitation properties of clouds. There is a heated argument amongst atmospheric scientists as to what will be the overall impact of warmer oceans pumping more water vapour into the atmosphere. If it leads only to more moisture in the lower atmosphere and heavier rainfall from clouds, its impact will be limited. If, however, it results in an increase in humidity throughout the troposphere, it will lead to a positive feedback producing greater warming because water vapour is the most important greenhouse gas (see Section 2.1.3). This issue has yet to be resolved.

9.2   SIMULATION OF CLIMATIC VARIABILITY

A fundamental test of the utility of GCMs in predicting climate change, is their ability to replicate this natural variability. This has been done by running GCMs for the equivalent of many centuries of the global climate without any external perturbations to see how much they go up and down of their own accord. On the shortest timescales they do a reasonable job of reflecting the synoptic variability of weather systems in the climate. They do, however, underestimate the variability of the regions of maximum storminess. On the other hand, the differences between the models are substantial and show more work needs to be done in this context. The models also tend to underestimate the amount of blocking (see Section 3.4) and, hence, the contribution this important factor makes to climate variability.

Comparison of the output of coupled models averaged over timescales of a month to a year suggest that the standard deviation of the near-surface temperature was in reasonable accord with observed values. The simulated values were generally larger than those observed in the tropics, but smaller in high southern latitudes.

Interannual variability is a more complex matter. Some of the atmosphere–ocean coupled GCMs display changes which resemble some aspects of the ENSO behaviour (see Section 3.5). Also, models which have simulated only ocean dynamics, produce realistic modes of interannual

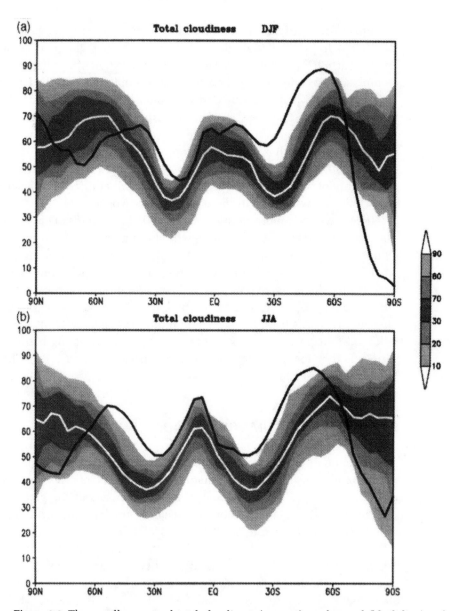

Figure 9.3 The zonally averaged total cloudiness (percent) as observed (black line) and as simulated by the international Atmospheric Model Intercomparison Project (AMIP) models for **(a)** December–February and **(b)** June–August. The mean of the models' results is given by the full white line, and the 10, 20, 30, 70, 80, and 90 percentiles data are given by the shading surrounding the model mean (IPCC, 1995, Fig. 5.10).

variability. The differences between the various models is, however, considerable, although when they are run using historic sea surface temperatures (see Section 6.1) they produce more realistic patterns of interannual variability which shows they are simulating important aspects of large-scale atmospheric–ocean interactions. But, the models cannot yet handle the seasonal cycle of the oceans and the associated surface heat and moisture fluxes, which is why models with flux adjustment do best.

When it comes to simulating decadal and longer fluctuations, there are considerable problems in establishing what constitutes the real level of natural variability. The combination of the limited geographical coverage of lengthy instrumental measurements and doubts about whether proxy records accurately reflect the scale of longer term variability mean the true level decadal and longer fluctuations have probably been underestimated (see for example Section 6.4.1). Nevertheless, because GCMs can be run under constant conditions for very lengthy periods, it is possible to get a measure of the variability predicted by models.

Two examples of long integrations of coupled atmosphere–ocean GCMs have explored the spectrum of multi-decadal to century timescale internally generated natural variability. These are a 1,000-year simulation on the GFDL model (see Table 9.1) and a 1,260-year run by the European Centre/Hamburg (ECHAM-1) model developed by the MPI. The results are summarised in Fig. 9.4, together with a shorter (310 years) run on the UKMO model (see Table 9.1 and Section 10.2). As can be seen the GFDL and UKMO models produce spectra which are quite close to the global mean temperature variability for periods up to about seventy years (the global temperature record is not long enough to draw conclusions on longer timescales). The substantial difference between the variance in the ECHAM and GFDL model runs is

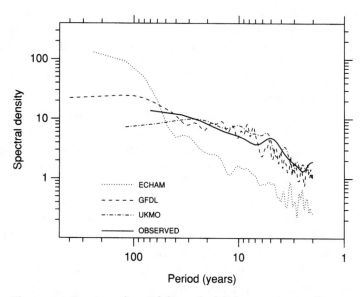

Figure 9.4 Spectra of variability of global mean, annually averaged near-surface temperature estimated from the observed data and from three ocean–atmosphere coupled model control runs. The observed data covers the period 1861–1994, while the model results are from simulations of between 310 to 1,000 years of global climate. The spectra provide information on the distribution of variance on timescales from two years to 400 years in the case of the longest model runs (IPCC, 1995, Fig. 8.1).

explained by an initial drift in the former. If this is excluded the difference is much less with the ECHAM variance flattening markedly for periods longer than 100 years.

These results suggest that the GFDL and UKMO models do a reasonable job of simulating the general level of variance in the global temperature for periods from two to seventy years. They do not, however, succeed in producing a peak at around three to five years which corresponds to the ENSO. The ECHAM model has the same problem with the ENSO and, more importantly, has consistently less variability over the range two to sixty years. So, on these timescales, the models are on the right track, but there are considerable differences between them, and their inability to represent the effects of the ENSO is a measure of their lack of detailed simulation of the real climate.

Another interesting feature of some of the advanced coupled atmosphere–ocean models (e.g. the GFDL model), is that they produce quasi-cyclic fluctuations on the timescale of several decades. This behaviour is similar to that observed in the climate. The cycle with a period of about twenty years may be an example of these interactive processes (see Section 8.1). Another example is a significant thirty- to forty-year variation of summer temperatures in Fennoscandia (Fig. 9.5) identified in the analysis of tree-ring width series (see Figs. 6.10 and 6.11). There is no *a priori* physical reason why this periodicity should occur (see Section 7.1). Furthermore, its frequency varies throughout the series. The fact that the GFDL model produces fluctuations in the North Atlantic on a similar timescale suggests that these variations in summer temperature could be the product of the internal variability of the ocean–atmosphere system. But this conclusion needs to be treated with caution given that, in spite of its sophistication, the GFDL model cannot simulate the Gulf Stream.

The limitations of GCMs in handling climatic variability become more obvious when it comes to considering changes on the longer terms and also regional variability. The tendency towards a flattening of the spectra from the ECHAM and GFDL models does not tally with the palaeoclimatic data obtained from proxy measurements (see Section 6.3), and the general property of the spectra of climatic variability to be 'red' (see Section 7.3). This flattening is probably a consequence of the 'flux adjustments' (see Section 9.1) made to keep GCMs on climatic straight and narrow.

An attempt to compare the variability of climate inferred from proxy measurements suggests that the amount of variability may be greater than previously assumed and that GCMs only reproduce between 20% and 60% of these fluctuations (Fig. 9.6). In this analysis the palaeodata temperature constructions from sixteen proxy records including tree rings, corals, ice melt records, ice cores and historical documentary records covering the period 1600 to 1950. So these records provide a measure of climatic variability at sixteen points around the globe. The model runs are two 350-year chunks of the ECHAM and GFDL results, shown in Fig. 9.4. To compare the climate variability inferred from palaeo-climatic data the output from the models is

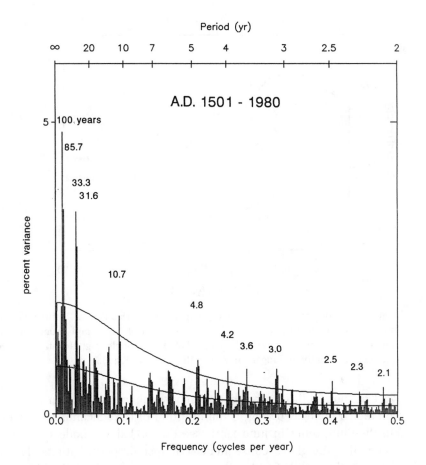

Figure 9.5 The variance spectrum of the reconstructed Fennoscandian temperature series (see Fig. 6.13) calculated over the period 1580–1980, showing that the distribution is 'red' and that the most significant feature is the periodicity at thirty-one to thirty-four years (with kind permission of K. Briffa, Climatic Research Unit, University of East Anglia).

calculated the sixteen points closest to where the proxy measurements were made. So this analysis is only capable of drawing regional conclusions and may not be relevant to global variability. Nevertheless, if the palaeoclimatic data is correct, it poses a major challenge for using model predictions to detect the impact of human activities. In short, the greater the scale of natural variability, the longer it will take to establish that any observed change can only be attributed to human activities (see Section 10.3).

An alternative means of checking how well GCMs deal with potential climate change is how they handle specific forms of natural perturbation. The best example of this process is the impact of large volcanic eruptions. Because these produce a reduction in the amount of solar radiation reaching the Earth's surface (see Section 8.4), which is of the same scale as the warming effect equivalent to a doubling of CO_2 in the atmosphere, the performance of models in computing the impact of major volcanoes is a good

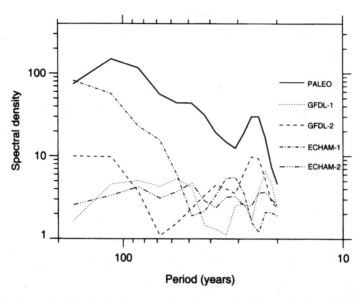

Figure 9.6 Spectra of near-surface temperature variability as estimated from two different ocean–atmosphere coupled model control runs and from palaeoclimatic temperature reconstructions, showing that the models significantly underestimate variance, especially on longer timescales (IPCC, 1955, Fig. 8.2).

test of their ability to predict the consequences of human activities. The predicted changes owing to the eruption of the volcano, Mount Pinatubo, in the Philippines in June 1991 (see Fig. 9.7) show models can do well in predicting the global consequences of specific perturbations. This success suggests that, while GCMs still have many limitations, their ability to predict the incremental consequences of human activities (see Section 10.2) makes them more valuable than their absolute performance might infer.

Another popular test of model performance is to explore how they simulate past climate. The most frequently examined are conditions during the Holocene optimum around 6,000 years ago (see Section 4.6) and the extreme opposite of the coldest period at the end of the last Ice Age at about 18,000 years ago (see Section 4.4). While the models produce what look like reasonable representations, the limitations in our knowledge of the climatic conditions of time make it difficult to establish whether such work is a real test of the models. Indeed, the benefit of these studies may be more to do with testing the physical implications of possible mechanisms of climate change against available evidence, than to validating the models.

9.3   THE CHALLENGES FACING MODELLERS

The potential to improve GCMs depends on making progress in a range of areas. There are obvious gains to be had in using bigger and faster computers. The most interesting developments are likely to take place, however, in

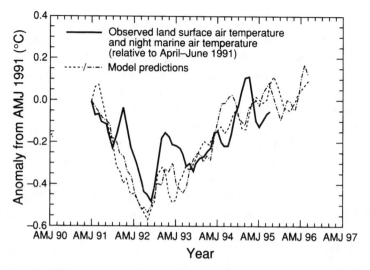

Figure 9.7 The predicted and observed changes in global land and ocean surface air temperature after the eruption of Mount Pinatubo, in terms of three-month running means from April–June 1991 to March–May 1995 (IPCC, 1995, Fig. 5.20).

better physical representations of the processes identified in Section 9.1. These include improved handling of clouds, the changing conditions of the land surface, such as soil moisture and snow cover, sea ice and the whole issue of 'flux adjustments'. Progress in these areas will depend on improved experimental studies of what is actually happening.

9.3.1 Clouds

Almost every aspect of the representation of clouds in GCMs requires improvement (see Section 9.1). At the most basic level the understanding of the physics of how much sunlight is absorbed and reflected by different types of cloud, and then how much heat radiation is absorbed and emitted by them must be quantified more accurately. Then the spatial and temporal distribution of different clouds must both be measured and then reproduced by GCMs. In addition, the processes controlling precipitation must be defined more accurately. Without progress on these fronts it will not be possible to provide more reliable estimates of how cloud cover varies over longer timescales and hence what part these variations play in climatic change.

The essence of the challenges facing modellers is found in the need to move away from the parameterization of clouds, to a more detailed treatment of the physical processes involved in the formation and behaviour of clouds. These improvements require better measurements of the radiative properties of water droplets and ice crystals both in respect of sunlight and terrestrial radiation for different types of clouds. An improved knowledge of the shape and size distribution of cloud particles is part of this understanding. In addition, measurements from the ground, and by aircraft and satellites of how single clouds and the combination of a number of clouds alter how radiation is absorbed, emitted or scattered by the atmosphere will be needed. Only when these are available will GCMs be able to produce more climatically realistic representations of current cloudiness.

Building up an improved climatology of cloud types using satellite measurements to identify how the amounts of various clouds vary on every timescale from diurnal to decadal is the next step. Once the geographical distribution of cloud types has been established accurately it can be used as a test for improving the performance of GCMs. Such measurements may also be used to check another potentially important aspect of human activities. This is the formation of particulates, notable from emissions of sulphur dioxide from the combustion of fossil fuels (see Section 8.8). What is needed is accurate figures on how much sunlight is absorbed and reflected by these particulates and how their net effect is modified by the albedo of the underlying surface.

In addition to the direct consequence of particulates, there is the indirect impact on cloud formation. This influences how much impact they have on the extent and duration of cloud cover. At present, modellers do not even know for certain whether these particulates will lead to an increase or decrease in cloudiness. Satellite measurements of ship trails show particulates increase cloud cover. So it is estimated that the net effect of particulates generated by human activities on the properties of clouds will probably be some additional cooling, but the amount cannot yet be quantified with any precision.

9.3.2 Land-Surface Processes

Many features of the land surface vary on short timescales. They do, however, exert a powerful influence on the climatic conditions where we all live. So, while it may be possible to parametrise many longer term aspects of the processes, it is essential that GCMs accurately represent their climatic consequences. As with clouds, the current evidence is that land-surface processes are not handled well by the models and the way forward is to obtain better measurements of how the properties of the land surface affect the behaviour of the atmosphere. These measurements will need to include changes in albedo (see Section 2.1.4) with the seasons and between wet and dry conditions, the balance between how much precipitation is either stored in the soil, or evaporates, or runs off in rivers, the impact of soil moisture on heat and moisture flux at the surface, and the consequences of winter snow cover.

The consequences of soil moisture variations are a particularly good example of how the challenges facing modellers are interlinked. Weather forecasting work (see Section 9.3.5) has shown that more accurate treatment of soil moisture can produce significant improvements in predicting rainfall. But in climate models, unless the general treatment of rainfall is realistic, weaknesses in the modelling of precipitation will overwhelm any improvements in the representation of land-surface processes. This basic challenge of identifying what are the most important parameters affecting a given climatic variable may be the biggest obstacle to progress. Nevertheless, only by exploring how the models respond to more realistic physical representations of land-surface processes may answers be found to some of the problems of simulating precipitation more accurately.

The impact of snow cover may produce more direct progress. Although the models do not handle global snow cover well (see Table 9.2), the scale of the effects of changes in the extent of snow at the end of the northern winter across the boreal forests is potentially large (see Section 8.9). These changes need to address the coupled issues of both how fluctuations in the climate may be amplified by parallel change in snow cover, and also how deforestation might alter these effects more radically.

The related issue of desertification must also be examined in more depth. The concern is that the high albedo of deserts may reinforce the dry conditions. But some theoretical analysis has suggested that the albedo increase owing to expanding deserts may be balanced out by a parallel decline in local cloudiness. More striking is the satellite observations which seem to show that the expansion and contraction of the deserts is dominated by larger scale weather patterns (see Section 8.9). So when rainy seasons return, the vegetation rapidly becomes re-established and reduces the albedo. If, however, longer term climatic change led to a more permanent shift in rainfall patterns then the albedo effects could be important. So the models need to be able to handle these changes.

9.3.3 Winds, Waves and Currents

The importance of atmosphere–ocean interactions in climatic change means an improved understanding of how energy is exchanged between the atmosphere and the sea surface is central to better modelling. The physics of how the winds stir up waves and are in turn influenced by the roughness of the sea requires improved analysis. Satellite measurements are now providing improved information about wave heights and wind speeds. These will contribute, together with oceanographic studies, to establish better figures for the amount of heat, momentum and moisture between the sea surface and the atmosphere as a function of the temperature contrast between the two and wind speed. This information will inform judgements about the handling of 'flux adjustments' in GCMs, especially as higher resolution models become available. Some of the most recent modelling work suggests that progress is being made to produce models that can make lengthy realistic simulations of the global climate without recourse to 'flux adjustments'.

On the larger scale, how the atmosphere and oceans combine to drive the ENSO and the ocean currents must be refined in the models. As current GCMs cannot reproduce the broad effects of the ENSO (see Section 9.2), this has to be a major priority in producing more realistic computer simulations of the climate. Beyond this, the processes which maintain the Great

Ocean Conveyor have to be modelled so that we can form a more informed view about the stability of this circulation system. Some preliminary model results suggest that increasing CO_2 levels may produce conditions which switch off the thermohaline circulation in the North Atlantic, especially if the levels rise rapidly. These results confirm the importance of producing improved models of ocean circulation.

9.3.4   Other Greenhouse Gases

The contribution to the greenhouse effect of other trace constituents of the atmosphere, apart from CO_2, has been recognised throughout this book. In addressing the impact of human activities it is important, however, to consider changes in all of the radiatively active gases. This analysis can be divided into two areas. The first group is the various anthropogenic emissions which will effectively accumulate in the atmosphere (e.g. methane, sulphur dioxide and CFCs). The second group is those constituents which respond in a complex manner to changing climatic conditions and so become an integral part of the modelling process, and water vapour and ozone are the most important. Although the former do participate in various atmospheric photochemical processes, their principal contribution to climate change is the amount they add to the radiative forcing of the atmosphere. This increase can be combined with the effect of CO_2 to provide an overall figure for human activities. By comparison, water vapour and ozone variations are much more challenging.

Because water vapour is the dominant greenhouse gas (see Section 2.1.3), its accurate representation is central to realistic modelling. This analysis must also recognise how sensitive the concentration of water vapour is to the temperature of the atmosphere, especially in the vertical. But the radiative properties of water vapour varies with atmospheric concentration and how the vertical distribution will be altered in a warmer world.

Similar issues arise with ozone. Here the vertical distribution is the product of photochemical processes (see Box 2.2). Human activities are altering this distribution. Near ground level rising pollution levels in urban areas are producing a widespread increase in O_3 levels especially in the summer. In the stratosphere, where the majority of O_3 is found, the build-up of CFCs has led to destruction of O_3, notably over Antarctica each austral spring. These changes must be adequately represented in GCMs to provide a realistic representation of how human activities will affect the climate.

9.3.5   Exploitation of Numerical Weather Prediction

At the beginning of this chapter, the scale of numerical weather prediction (NWP) was described to underline how, in spite of developing massive computer models of the global weather system, the forecasts have profound limitations. Progress in improving day-to-day weather forecasts also has direct relevance to the challenges facing climate modellers. This is that it can provide rapid insights into which refinements to the physical representation of the climate system have the most impact on performance. By using the latest experimental observations of how atmosphere interacts with the oceans and the land surface, weather forecasting models can improve the representation of various physical processes. The impact of these refinements can be checked in terms of improvements in forecasting performance. This process offers climate modellers the prospect of being able to identify which factors are likely to have the most impact on their longer term predictions and how best to represent them in GCMs.

9.4   SUMMARY

Climate modelling has many features of the meeting of an irresistible force and an immovable object. The computational apparatus available to modellers is massive, and when compared with many aspects of managing complexity in our lives, the analysis it provides is far more comprehensive. Furthermore, our understanding of the physical processes involved has come on with leaps and bounds. Nevertheless, the realisation of just how enormous a challenge the production of a reliable representation of the global climate is, has grown as fast, if not faster.

   Confronted with what looks like an intractable problem, it would be all too easy to throw our hands up in the air and declare it is all too difficult, and that nothing can be concluded until we have better measurements, bigger computers and more sophisticated GCMs. But this extreme 'wait and see' approach is not realistic. We have to make decisions now. Even taking no action to respond to the threat of global warming amounts to a decision and has to be justified in terms of current knowledge. The only way forward is to exploit the work that has been done while recognising its limitations when deciding what are the most worthwhile things to do. So, having identified the strengths and weaknesses of modelling the climate we can now consider how this work can be used to plan the future.

QUESTIONS

1   The ocean–atmosphere models show a variation in poleward water transport in the North Atlantic of 2 to 26 Sv (1 Sv = 10^6 m^3 s^{-1}). If this water has an average temperature of 11 °C when it arrives at high latitudes whereas the deep water returning southwards has a temperature of 3 °C, estimate the range in the amount of heat being transported northwards in these models. How does this variation in energy compare with the amount of solar energy falling on the region (50–60° N and 20–60° W) in winter and summer?

2   The difference in the figures in Tables 9.1, 9.2 and 9.3 (e.g. global temperature, precipitation and extent of snow cover and sea-ice, etc.) produced by

various GCMs are much greater than the predicted changes for the coming century. Does this invalidate any conclusions reached about future climate change using these GCMs and, if not, why not?

FURTHER READING

A complete reference list is available at the end of the book but the following is a selection of the best books or articles to follow up particular topics within this chapter. Full details of each reference are to be found in the Bibliography.

Burroughs (1991). This provides descriptions of how satellite technology is being used to obtain better measurements of how various physical processes are contributing to climate change.

Burroughs (1994). This contains extensive discussion of how the various components of the climate may interact with one another to produce cyclic, or quasi-cyclic behaviour.

Gurney et al. (1993). A thorough presentation of many of the results that have been obtained from weather satellites which are of relevance to climatic studies.

Harries (1990). A more fundamental analysis of how weather satellites work, and a guide to the contribution they can make to studying climate change.

IPCC (1995). The definitive statement on where the meteorological and climatological community have got to by the end of 1995 in measuring climate change and modelling the climate. The analysis of the performance of GCMs provides the ideal start for finding out more about where this rapidly advancing subject has got to and the challenges it now faces.

Trenberth (1992). A thorough introduction to the physical and computational principles that form the basis for modelling the global climate.

CHAPTER TEN

PREDICTING CLIMATE CHANGE

Some people ask, "What if the sky were to fall?"

Terence (Publius Terentius Afer) c190–150 BC

It follows from the discussion of computer models of the climate in Chapter 9, that predicting how the climate may change over the next century or so can be divided into two areas. The first is to decide how much is known about natural fluctuations in the climate and what this means for the future. The second is the whole question of how human activities will develop and what impact these will have on the climate. These issues are best considered separately and then, in the light of any conclusions reached, combined. In doing so, the objective will be to draw on all the material in this book to sum up how our knowledge on climate variability and climate change can be used to plan for the future.

10.1 NATURAL VARIABILITY

The relative stability of the climate throughout the Holocene poses a problem when considering how the climate may change over the next hundred years. It is all a matter of whether we can regard this stability as being the current natural order of things. If so, then an accurate measure of fluctuations over this period defines the current natural variability of the climate. If not, then we may have to include some elements of climate change, which occurred prior to the Holocene, in the analysis. In short, we must decide which version of natural fluctuations is the best choice before we can decide whether human activities are having a significant impact.

The absence of a well-defined trend over the last few millennia means that we can start with periodic change. Although there is a strong case for orbital variations being the pacemaker for the ice ages, these will only exert an influence on longer timescales. One estimate is that we are headed for another ice age in about 23,000 years time. More immediately, there is very little evidence of regular changes on shorter timescales (i.e. from a few years to a few millennia). Apart from the ubiquitous nature of the 'twenty-year cycle' which has variously been attributed to solar, lunar or atmosphere–ocean autovariance, there is less substantial evidence of cycles around 100 and 200 years, which may also be linked to solar activity (see Section 8.5). There is also growing evidence that there may be some quasi-cyclic variation of between one and two millennia in duration. But, none of these features is sufficiently well-established, or well-understood to be part of our forecasts for the twenty-first century.

This leaves the issue of more sudden and sizeable shifts. Here the issue is whether using our knowledge of climate change during the Holocene under-estimates the capacity of the global climate, in its current form, to change more radically. Compared to events prior to around 10,000 years ago, we may be living on borrowed time (see Fig. 4.13). But it is in the nature of the non-linear behaviour of the global climate that we have no way of telling whether it will remain this stable, or when it might flip into a new state. All we can say is that, at a time when human activities are predicted to have an appreciable impact, the chances of the climate becoming more unstable will increase. At the same time, there is always the possibility of a sudden natural change occurring, which has nothing to do with human activities.

The implications of these conclusions are to undermine confidence in any predictions of future natural climatic variability. To the extent that there are any 'cycles', their impact is likely to be small, and difficult to quantify until an accepted physical cause is identified. This places particular emphasis on finding out whether solar activity plays a part in these changes, given the cyclic variation in sunspot numbers and the close parallel between their longer term behaviour and global temperature trends.

The question of whether the climate could shift to a more variable pattern as a consequence of natural changes, rather than as a result of human activities, remains in the realms of speculation. Progress depends on coupled atmosphere–ocean models reaching a degree of sophistication where we have confidence in the changes they may predict in, say, oceanic thermohaline circulation. This will require modellers to overcome fully the challenges set out in Section 9.3. In the meantime, the priority must be to use all the available sources of palaeoclimatic information to generate a better understanding of natural variability and what triggers more dramatic shifts in the climate. Without this, it will take far longer to decide whether future changes are part of the predicted impact of human activities, and hence in theory avoidable, or simply part of the normal ups and downs of the climate.

If, however, the climate does undergo a sudden shift, it will be entirely academic as to whether it was a result of natural variability or human activities, as it will be too late to do anything about it.

10.2   PREDICTING GLOBAL WARMING

In spite of the limitations in the capacity of general circulation models (GCMs) to reproduce all the details of the current global climate described in Chapter 9, they are the only physically realistic way to predict the impact of human activities on the climate. Furthermore, although their predictions cover a considerable range of climatic conditions, they are consistent in their broad prediction. This is that the radiative forcing due to the build-up of greenhouse gases in the atmosphere (see Section 2.1.4) will exert a dominant influence on the future global climate change. The consensus figure adopted by the IPCC is that, in terms of greenhouse gases, the radiative equivalent of a doubling of the concentration of CO_2 in the atmosphere (from 275 to 550 ppm) will lead to a rise of between 1.5 and 4.5 °C in the mean global temperature, with a best estimate of 2.5 °C.

Because the various GCMs adopt such a wide range of approaches to handling the many challenges in simulating the global climate, it is not practical to present a review of the different predictions they produce. Instead, it is more illuminating to consider a single example which is typical of the current status of climatic modelling. This is the work of the Hadley Centre at the UK Meteorological Office, at Bracknell, England. A recent set of results gives a good idea of the scale of the computations required to form a reasonable picture and the limitations of the results obtained. It also provides a measure of the latest thinking on the consequences of doubling the radiative forcing of CO_2 in the atmosphere, and also includes the effect of sulphate aerosols created by the combustion of the fossil fuels containing sulphur. The latter is particularly important as it may offer an explanation as to why the warming during this century, with its temporary abatement between the 1940s and the 1970s (Fig. 10.1), has not followed the course predicted on the basis of the build-up of greenhouse gases alone.

The model was brought to near equilibrium through 470 years of simulated climate. After that it was run for 300 model years under control conditions during which there was no detectable trend in the global mean temperature. Thereafter, three experiments were conducted each starting with the conditions in the year 1860 and running up to the year 2050. These consisted of a control with constant CO_2 concentrations, an experiment in which the concentration of CO_2 is increased gradually to give changes in radiative forcing due to all greenhouse gases, and an experiment in which both greenhouse gases and the direct radiative effect of sulphate aerosols was represented. Possible indirect effects of sulphate aerosols on cloudiness (see Section 9.3.1) are not included.

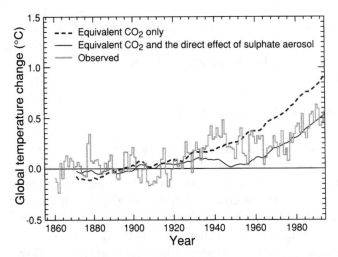

Figure 10.1 Simulated global annual mean warming from 1860 to 1990, allowing for increases in equivalent CO_2 only (dashed curve) and allowing for increases in equivalent CO_2 and the direct effect of sulphate aerosol (flecked curve), compared with observed warming that has occurred over this period (stepped grey curve). The anomalies are relative to the 1880–1920 mean (IPCC, 1995, Fig. 6.3).

In the model the radiative forcing due to greenhouse gases rises to 2.5 W m^{-2} by 1990 and thereafter at a constant rate of 0.6 W m^{-2} per decade. The mean global value of radiative forcing by sulphate aerosols is put at −0.6 W m^{-2} in 1990 and this increases to about −1.2 W m^{-2} by 2050. The effect of this cooling is to reduce future warming from 0.3 °C per decade to 0.2 °C per decade. But the effect of sulphate aerosols reflects the forecasts of regional changes in industrial activity in the next century. It is greatest over India and China, and also significant over North America and much of Eurasia.

These results have a number of important implications. First, the inclusion of sulphate aerosols has the important effect of producing a more realistic variation in global temperature between 1860 and 1990 (see Fig. 10.1). This bolsters confidence in the prediction. Secondly, it reduces the value of the warming by the time greenhouse gases have built up to the equivalent of a doubling in CO_2 from 2.5 to 1.7 °C. After 2050, the model shows that greenhouse gases get the upper hand and by 2100 the warming is expected to be about 3 °C above pre-industrial levels. The third important consequence is that the pattern of warming tends to mirror some of the changes observed in recent decades with less warming over the northern continents. This suggests that the inclusion of sulphates produce what look like more realistic forecasts but also that future warming could exhibit significant regional variations.

The success of introducing sulphate aerosols into the GCMs does, however, show that the inclusion of new factors can have an appreciable impact on predictions of future warming. Given the issues raised in Section 9.3, it is possible including other factors (e.g. the indirect impact of sulphate aerosols on cloudiness, dust generation, desertification and deforestation) in the models, could alter the predicted warming appreciably. Nevertheless, the fact remains that the consensus view of the meteorological and climatological community is that human activities will on balance lead to a warming of the global climate. So, by the end of the next century the temperature could be about 2 to 3 °C above current levels unless appreciable action is taken to cut back on the predicted emissions of greenhouse gases.

The warming is predicted to be greater at high latitudes, especially in the northern hemisphere, and somewhat less in the tropics. This is consistent with the standard view of past climate change, notably the ice ages, which has concluded that the tropics are more stable than higher latitudes. Recent studies suggest that this may understate the sensitivity of the tropics to major shifts in the systems controlling the climate, and this will need to be studied closely in testing the performance of GCMs.

This consensus does not get away from the uncertainties of Section 10.1. Because much of the warming over the last century may have been the product of the natural variability of the climate, it could be purely coincidental that the models produce a reasonable fit. In spite of the recognised limitations of the models, this conclusion is probably uncharitable for at least one good reason. This is the fact that the models are not 'tuned' to produce a good fit with observed temperature trends, but are based on sound physical principles. This means that once constructed and kicked off with a realistic set of starting conditions they run on their own accord to simulate how the world should have responded with and without the atmosphere being modified by human activities. The fact that the modelled world looks rather like the world we live in, and the changes induced by human activities look similar to events in recent decades is good reason for taking these forecasts seriously.

10.3 THE PREDICTED CONSEQUENCES OF GLOBAL WARMING

If confidence in models is justified, the next question is what are the implications for our lives if the global temperature rises by around 3 °C by the end of the twenty-first century? The answer to this question depends on three principal issues:

1. how regional weather patterns are altered by global warming;
2. whether the weather will become more extreme; and
3. the rise in sea level.

The analysis of the performance of GCMs in Section 9.1 notes the scatter between the output of different models. It is hardly surprising, therefore, to discover that various simulations of future regional weather patterns in a warmer world show a wide range of possible developments. These differences reflect not only the physical components of the model (e.g. whether or not flux adjustments are included), but also which human activities are considered, especially those which will vary regionally in their impact (e.g. particulates). For instance, the UKMO model suggests that the incorporation of sulphate particulates significantly alters the simulation of the impact of global warming on the Indian monsoon. Without particulates the model, along with most GCMs, predicts an increase in rainfall. This behaviour is consistent with a general conclusion that increased temperatures in the tropics will mean more water vapour in the atmosphere and hence stronger hydrological cycle. When particulates are included the UKMO model produces drier conditions over India.

The same uncertainties apply when considering how warming will affect higher latitudes. The cosy notion that warming will simply produce a gradual displacement of climatic zones to higher latitudes, so England would eventually have a climate like southern France, is probably a gross oversimplification. What will matter more is whether, instead of a general warming, there is a significant shift in the incidence of weather regimes (see Section 7.3). As noted in Section 9.2, GCMs cannot handle the changing behaviour of phenomena such as blocking yet. So, while mid-latitudes are likely, on average, to become warmer there could be significant regional and seasonal variations.

The prediction of whether the weather will become more extreme is another facet of the question of the incidence of regimes. The hydrological cycle will probably become stronger. This is likely to increase the frequency of extreme rainfall events. The incidence of very low temperatures can also be expected to fall. As for the other extremes, which are part of the popular perception of global warming (e.g. stronger and more frequent hurricanes and mid-latitude storms), the results from GCMs are equivocal. Higher tropical sea surface temperatures could spawn more hurricanes although this is not substantiated by the models. Moreover, there is little evidence of any increase in hurricane activity associated with the warming during the twentieth century (Fig. 10.2). Indeed the decline in the tropical Atlantic has been linked with the incidence of stronger El Niño events in recent decades, as warm conditions in the eastern Pacific dampen down tropical storms in the Atlantic. What has yet to be established is whether more frequent El Niño events will be part of the warmer world predicted by the models.

In the case of extratropical depressions, the models tend to predict a reduction in the intensity of these storms and of the incidence of gales over Europe. This reflects the fact that if the polar regions warm more than the tropics then temperature gradient driving these storms will be reduced and the storm tracks will move to higher latitudes. If the strength of mid-latitude circulation declines, especially in winter, then it could lead to more meandering weather patterns and a greater incidence of blocking. One consequence of this change would be more frequent cold winters in

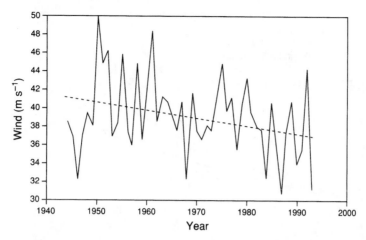

Figure 10.2 Time series of mean annual maximum sustained wind speed attained in Atlantic hurricanes between 1944 and 1993. The linear trend is shown as a dashed line (IPCC, 1995, Fig. 3.19).

some parts of the northern hemisphere, especially in the early stages of the warming. This change might be matched by hotter summers. The impact of these shifts on rainfall patterns is not yet clear, and the inconclusive figures on current trends, apart from the drier conditions in the Sahel, provide few clues to future patterns.

The consequences of shifting regional patterns and greater climate variability are likely to impact most on agriculture. While seasonal forecasting may alleviate the worst effects of these fluctuations on some annual crops, the role of government in enabling farmers to survive from year to year will play a major part in responding to these challenges. The other economic consequences will depend crucially on the nature of the increased variability. Here the scale of the impact will depend on what action public and private institutions take to influence on the economic decisions we all make. For instance, the extent to which coastal areas vulnerable to increased hurricane activity are developed will depend on whether local authorities tighten up planning regulations and whether insurance companies are willing to provide adequate cover at economically acceptable rates.

The related issue of rising in sea levels is governed by the same processes. In climatological terms the rate of rise is linked to precipitation patterns at high latitudes. If warming leads to increased precipitation over Antarctica and Greenland then any greater peripheral melting of the ice sheets will be more than balanced by the accumulation of snowfall at higher levels. At lower latitudes, where glaciers and ice caps have been receding since the late nineteenth century, warming has clearly outweighed any increase in precipitation. So, for much of the coming century the rise in sea level will be dominated by the thermal expansion of the oceans plus the melting of ice caps and mountain glaciers at lower latitudes. The melting of the ice sheets of

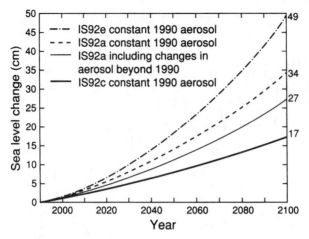

Figure 10.3 Model projections of global sea level rise over the period 1990–2100 assuming that future warming is moderated by the impact of sulphate aerosols. The various curves reflect differing assumptions about the levels of greenhouse gas emissions over the period (IPCC, 1995, Fig. 7.10 [without the labelling]).

Greenland and Antarctica is only to become a major problem in the longer term (see later). The latest estimate of the median rise in sea level is 27 cm by 2100 with a range from 17 to 49 cm (Fig. 10.3). This calculation assumes a relatively conservative climate sensitivity of 2.2 °C for the equivalent of a doubling in CO_2 levels, which reflects the possible impact of aerosols. This compares with the figure of 10–25 cm over the past 100 years (see Section 5.4).

The choice of a conservative estimate of future sea level rise is because this is an area where there has been a marked reduction in forecasts over the last decade or so. This cautious approach does not, however, represent complacency for two important reasons. First, the range of predicted changes is so dependent on forecasts of future fossil fuel consumption that inevitably any figure has to be treated with care. Second, and more significant, is that even the low figure for the rise by 2100 has a powerful message. This is that once the process of thermal expansion, which is the cause of most of this sea level rise, is set in motion it is likely to continue for centuries. So even the lowest predictions imply that in due course the sea level will rise to a damaging extent: it is only a matter of time.

How a rise in sea level will affect different parts of the world also depends on how the Earth's crust will adjust to past and future changes in the load of the major ice sheets. This glacial isostatic adjustment means that the impact will vary from one coastal site to another. Superimposed on this post-glacial rebound will be regional isostatic and tectonic effects. In addition, local factors such as groundwater extraction and land reclamation, further complicate matters. So analysis of the specific impact of a rise in sea level will depend on a combination of improved global models of post-glacial rebound and local geological knowledge.

In spite of all these uncertainties, the actual progress of sea level rises is likely to be an inexorable process which can be monitored closely. In the short term its impact will be dominated by the increasing vulnerability of low-lying coastal areas to severe storms. So the overall social response to these events will be an extension of the institutional reactions identified above. Local and regional governments will have to reach long-term decisions about how much is invested in coastal defences as opposed to allowing some land to be lost to the sea. At the same time, individuals will have to make decisions about the risks they are prepared to take in respect of themselves and their property. The extent to which they can get adequate insurance cover is bound to influence these decisions. In the longer term, of course, mounting costs will provide ever-greater incentives for central governments to make more substantial and lengthy provisions to prevent damage rising to politically unacceptable levels.

The more dramatic scenarios of the melting or collapse of the Greenland and Antarctic ice sheets appear to be distant prospects because of the likely effects of increased precipitation over the ice sheets in the next century or so. But uncertainty surrounds the stability of the West Antarctic ice sheet, which rests on a rock bed well below sea level. Depending on the physical assumptions made about the conditions under the ice, theoretical models can produce equivocal predictions about whether this ice sheet could collapse catastrophically if global warming and the sea level rose to some threshold value. This prospect has been the subject of a large number of studies and the overall judgement is that possible scenarios range from the Ross Ice Shelf disintegrating over the next 200 years and the West Antarctic Ice Sheet collapsing rapidly over 50–200 years raising sea level 60–120 cm per century, to snowfall increasing and the Antarctic ice sheet making a negative contribution to the sea level until global temperature has risen at least 8 °C, something that is not predicted to happen until 2200 at the earliest.

Whatever the actual response of the ice sheet, the fact of the matter is that both the timescale of any change and scale of the impact of any collapse far too big to be the subject of a plans now. The only solution is to take action in good time to prevent global warming reaching a level which makes such an event likely. So, everything depends on establishing as soon as possible the validity of the forecasts of future warming, and, in particular, any predictions of sudden changes in major components of the climate system.

10.4 WHEN WILL WE BE CERTAIN ABOUT GLOBAL WARMING?

Those involved in assessing the contribution of human activities to global warming fall into two principal camps. The majority regard its impact is already a proven fact. The minority consider the evidence as inadequate and its reality will only be established when a far more sustained warming has been observed. The median view is the guarded consensus reached by the Intergovernmental Panel on Climate Change (IPCC) in 1995 that 'The balance of evidence suggests a discernible human influence on global climate through emissions of carbon dioxide and other greenhouse gases.' The differences in opinion about this centrist view reflect the uncertainties in defining the natural variability of the climate and modelling its behaviour. Equally important is the interests of the various parties who stand to gain or lose as a result of any action taken to prevent future predicted warming. This conflict will not be resolved until the match between predicted and observed changes is close enough to meet certain criteria. This is a matter of statistical analysis and is often termed *fingerprinting*.

The essence of fingerprinting is that in some way the impact of human activities will be appreciably different to natural variability. If this is the case then with a sufficiently accurate model it will be possible to define how, say, the spatial distribution of global atmospheric temperatures will change as a result of human activities. Then a statistical analysis of how much of the observed variance can be attributed to the human activities will establish how much confidence one can have in the forecasts. If the observed patterns move farther away from the climatic normal in a manner consistent with the model predictions, then an increasing proportion of the observed variance can be put down to us. At some stage the case for the human fingerprint on the climate will become incontrovertible.

Early efforts at fingerprinting are central to the IPCC conclusion on the discernible nature of the impact of the build-up of greenhouse gases on the global climate. These studies have concentrated on the three-dimensional pattern of temperature change and how this might reflect human rather than natural factors. Two particular features have attracted most attention. First, the regional distribution of temperature change in recent decades (see Fig. 4.23) and how this pattern is more accurately reproduced by models which include the effects of sulphate aerosols (see Section 10.2). The second is how the stratosphere has cooled while the lower troposphere has warmed (see Section 6.2). Here again the observed changes are consistent with the predictions the GCMs produce for the consequences of the build-up of greenhouse gases and with the basic physics of the greenhouse effect (see Box 2.1). Both these developments support the thesis that the observed trends are increasingly difficult to explain away without including human activities.

On the other hand, much of the variance in surface temperature observed in recent decades is attributable to the ENSO fluctuations and the North Atlantic Oscillation. But computer models have yet to reproduce the most obvious features of the variation in the incidence of either these broad climatic fluctuations, or the various attendant weather regimes (see Section 3.2) that are a major part of the process of observed climate change in recent decades. So, there is the sneaking suspicion that a lot of what is being observed, while difficult to explain without including human activities, is due to natural variability.

The emerging consensus on the causes of the current warming underline the strengths and weaknesses of using GCMs to detect the impact of human activities. The models deal only with the most obvious physical effects and, as discussed in Section 9.3, there are a variety of other processes which could play a significant part. It is not just natural variability we have to be worried about. At the very least, the indirect effects of particulates on cloudiness, the effects of changes in other greenhouse gases, changes in the albedo of land surfaces and solar variability need to be included in the models. The 1995 IPCC report provides a clear discussion of the issues involved in the inclusion of an increasing number of factors whose impact on radiative forcing may produce a better fit between model predictions and observed temperature trends. The scale of the uncertainties mount as we move beyond the basic contribution of additional greenhouse gases (Fig. 10.4). Depending on the assumptions made about the impact of parti-

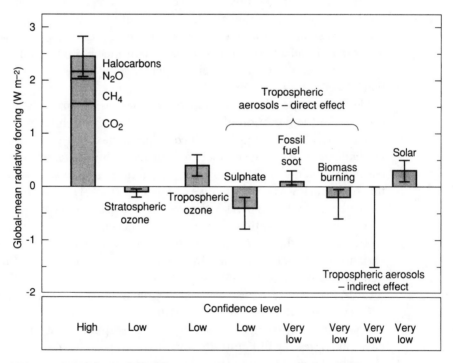

Figure 10.4 Estimates of the globally and annually averaged anthropogenic radiative forcing (in W m^{-2}) due to changes in concentrations of greenhouse gases and aerosols from pre-industrial times to the present day and to natural changes in solar output from 1850 to the present day. The height of the rectangular bar indicates a mid-range estimate of the forcing whilst the error bars show an estimate of the uncertainty range, based largely on the spread in published values. At the bottom is a measure of the confidence of the IPCC that the actual value of the forcing lies within the indicated error bars (IPCC, 1995, Fig. 2.16).

culates, volcanoes and solar activity, the quality of the fit can be improved, but at the cost of introducing ever greater uncertainty about just how important is the contribution of rising CO_2 levels.

One interpretation of this analysis is that the combination of the central figure for the sensitivity to increasing greenhouse gases equivalent to doubling CO_2 levels, together with reasonable estimates of aerosol, volcanic and solar effects produces an excellent fit of past trends. A more sceptical view is that increasingly we run the risk of falling into the trap of 'tuning' our models to come up with a better fit. This means each time the climate fails to behave as predicted the sensitivities will be adjusted, or additional factors introduced to obtain a new fit. So while the physical credentials of the models remain sound, the temptation to select values for the human perturbations, from within an acceptable range of uncertainty, which fit current observations, could lead to a major distortion of our judgement.

In the case of stratospheric cooling, other factors intervene in a different manner. Because of the decline of ozone (O_3) levels in recent years, notably at high latitudes, the amount of solar radiation absorbed in the stratosphere has been reduced (see Box 2.2). This has led directly to a cooling at these levels. This change, although almost certainly due to human activities, is physically separate from the analysis of the build-up of greenhouse gases in the troposphere. So the models need to consider both how changes in O_3 alter both solar absorption and also affect outgoing terrestrial radiation given that O_3 is a powerful greenhouse gas. Results on the GISS model (see Table 9.1) suggest that inclusion of O_3 changes, if anything, strengthens the case for the detection of the climatic fingerprint of human activities.

The real test of the models will come in the next few years. If the current predictions are broadly correct the impact of human activities will soon become indisputable. The warmest year on record since 1860 was 1997, but this high mark was comfortably exceeded by 1998. The five warmest years have all been in the 1990s. But, the 1997 and 1998 record levels were largely a reflection of the exceptionally warm El Niño event during much of the time, and this raises the question as to what extent this upsurge was a natural variation. As noted earlier, whether there will be more warm El Niño events in a warmer world is still the subject of scientific debate. Nonetheless, recent record-breaking warm years suggest that the models are on the right track.

This is all well and good, but what if the upward trend in surface temperatures experiences a significant interruption? The modellers will have much greater difficulty in convincing those who have to make decisions about global warming that their actions should be based on the output of GCMs. It will not be realistic to assume that producing an additional factor, like a rabbit from a hat, just as particulates may have provided an explanation for the interruption in the warming trend between around 1940 and 1970, will satisfy potential users. The power of the arguments for taking action about global warming is that the principal cause is the emission of greenhouse gases, and that these emissions can be reduced without politically unacceptable social and economic costs. If, contrary to current trends, reality turns out to be a much more subtle balance of warming and cooling effects blurred by greater natural variability than previously assumed, then the case for specific action is more difficult to make. In this situation, tackling climate change will slip down the list of priorities. It will become one of the many factors that have to be considered in making difficult decisions democratic societies make about environmental and social conditions.

This conclusion means that until there is unequivocal evidence that much of the observed global warming is due to build-up of greenhouse gases, any action to curb their emissions should not be driven by climatic arguments alone. This does not mean we should do nothing. Instead, as with so many other political decisions, the analysis must take full account of the conflicting economic, environmental and social issues involved. This process recognises that there are other priorities (e.g. education, health, and law and order) which may place more urgent demands on public spending. Given these pressures, it does not help to gloss over the limitations in our current understanding of climate change. To do so and argue that there is a simple answer, runs the risk of being accused of not having understood the question.

10.5 CAN WE DO ANYTHING ABOUT CLIMATE CHANGE?

These cautious observations about the challenges facing governments in reaching decisions on what action should be done to minimise the threat of climate change looks like a cop out. Given the scale of the work that has been done by the IPCC, surely there must be sufficient agreement about what should be done. In particular, the international understandings, which have emerged from major UN-sponsored conferences like those held in Rio de Janeiro in 1992, Berlin in 1995 and Kyoto in 1997, suggest that in terms of curbing the build-up of greenhouse gases in the atmosphere, there is a measure of agreement on what action should be taken. A closer look at both what the international community has agreed suggests otherwise. Limited real action, and the bitter arguments swirling around the negotiations shows, however, that the achievement of lasting reductions in greenhouse gas emissions will involve more than setting targets.

It is probably best to leave the basic arguments about how the consumption of fossil fuels will rise during the twenty-first century on one side. This may seem surprising, but there have been considerable fluctuations in the rate of growth of CO_2 levels since the mid-1970s. Apart from fundamental questions about the rate of uptake of CO_2 by the biosphere, there are huge doubts about the future rate of growth of the global economy and how CO_2 emissions will be linked to this growth. In the late 1990s there seems to have been a sharp decline in these emissions while the global economy has been growing rapidly.

How the burden of reducing emissions is apportioned between the developed and developing nations is another tricky political issue. Trading emissions between countries will almost certainly have a part to play, but whether it can become a central plank in future strategies depends on how it impacts upon individual countries. Negotiations to achieve workable schemes are bound to continue for many years. They will be coloured by how the climate actually behaves and shifting perceptions about whether other human activities are making important contributions to the observed changes. The going will get tougher as the economically attractive and politically acceptable options are used up. So advanced industrial nations will be willing to reduce CO_2 emissions by sensible energy conservation measures, improved public transport in major urban areas, and closing down ageing coal-fired power stations. But when it comes to the imposition of unpopular taxes on energy consumption or the restriction of the use of private vehicles, their resolve to meet demanding targets may weaken. At the same time, developing countries, whose per capita energy consumption is much lower, will continue to argue that they should be allowed leeway to achieve economic growth. If these complications are combined with confusing developments about other factors contributing to climate change the negotiations will become increasingly muddied.

All of this means there is huge uncertainty about what action can be taken to minimise the impact of human activities on the climate. For this reason the many scenarios of the future build-up of greenhouse gases in the atmosphere together with the forecasts of how these figures would be reduced by the adoption of certain strategies are not reviewed here. Suffice it to say they cover a wide range. Depending on future economic growth and the action to reduce emissions, the level of CO_2 in the atmosphere could either rise to a level well over 800 ppm by the end of the twenty-first century or be stabilised at 500 ppm by around this time. This would involve anthropogenic CO_2 emissions either more than doubling, or being reduced to barely a third of current levels by the year 2100. The essential feature of these figures is that whatever action is taken now to reduce emissions, the levels of greenhouse gases in the atmosphere are bound to rise over the next century or so.

Depending on what is done and how quickly it is implemented, the rate of growth will be slowed, but the consequences will only become appreciable in the longer term. The crux of the matter is how fast the climate reacts to this inevitable build-up of greenhouse gases. So we have to hope that the low sensitivity estimates (i.e. the temperature rises by only 1.5 °C for a doubling of CO_2) are nearer the mark, and that any warming is insufficient to trigger any more radical climate shift within the foreseeable future. This would then give us the maximum time to put our house in order. If, however, the climate warms up more rapidly, or reaches some critical level which precipitates a flip to some new climatic state, then it could well be too late to do anything about it.

Unfolding weather events will be a central role in political perceptions of what needs to be done. These will vary from country to country and governments will only take more radical action to reduce emissions when this is seen as essential to address pressing needs. So, concerted action will not be taken until there is incontrovertible evidence of damaging changes in the climate, by which time many of these shifts are bound to run their course. This means that improving our knowledge of what is really driving current climate change is the only way of planning for future developments.

One way of thinking about this difficult issue of how to deal with uncertainty is to consider the consequences of a sudden and dramatic shift in the climate. For instance, as discussed in Sections 3.7 and 8.3, the circulation in the North Atlantic is capable of switching into different modes. If it were to reverse the change experienced at the end of the Younger Dryas, 12,000 years ago, the impact on winter temperatures across much of the northern hemisphere would be catastrophic. They would drop by several degrees Celsius, especially in northern Europe, and the economic consequences would be unimaginable. Given such an event has not happened for 12,000 years, and is by its very nature unpredictable, it is not realistic for governments to press ahead with costly policies to prevent the climate flipping. All they can do is stick to those policies which appear worthwhile, whether or not the climate is going to throw a wobbly. These are bound to concentrate on actions which make economic sense now; may well delay global warming; and could conceivably have the additional benefit of reducing the risk of more catastrophic events occurring, albeit only by a tiny amount.

10.6 THE GAIA HYPOTHESIS

Having explored the range of explanations of climate change, and what we can do about it, and come up with what can only be described as an equivocal, if not gloomy, set of conclusions, you may well ask whether there is a more convincing way of presenting the arguments. I would insist that this is not the case, but you could respond by saying 'I would say that wouldn't I'. So as a closing presentation or even an epilogue, I will leave you with what could either be regarded as an alternative explanation or simply another way of looking at the evidence. This is the controversial 'Gaia hypothesis'.

This hypothesis was first proposed by James Lovelock, an independent English scientist and named after the 'mother Earth' goddess of the ancient Greeks. It aims to explain why, unlike other planets in the solar system, the Earth's atmospheric composition and history cannot be described in terms of physics and chemistry, but reflects the strong influence of biology. It explains the survival of life on Earth for nearly four billion years by treating life and the global environment as two parts of a single system. In effect, micro-organisms, plants and animals behave in such a way that the Earth's environment becomes adjusted to states optimum to their maintenance. This is not a conscious act on the part of the biosphere but instead that adjustments arise from natural selection. As such it represents a more sophisticated interpretation of the biosphere than has been considered earlier in this book (see Section 3.5).

A good example of the type of process that might contribute to the stability of the global ecosystem is the production of dimethylsulphide (DMS) by phytoplankton. As noted in Section 3.5 this gas is the biproduct of the life-cycle of marine algae. Its conversion into sulphate particulates may be a major factor in the formation of clouds over the oceans (see Section 9.3.1). So it is possible that an increase in sea surface temperatures together with a rise in CO_2 levels will lead to a greater production of algae and hence release more DMS into the atmosphere. If this results in additional cloud formation this will have a cooling effect – a negative feedback mechanism which effectively acts as a climatic 'thermostat'.

It is not necessary to go further into the competing arguments that have swirled around the Gaia hypothesis since it was first published in 1972, or the various physical processes and models that have been developed to explore its physical validity. What matters here is the concept of the biosphere, in responding to the types of physical changes which can cause climate change (see Chapter 8), meets the criterion of being *optimum for the maintenance of living things*. While there is a heated debate about what this criterion means, in the case of many of the examples cited in this book it can be used to consider the overall response of the climate system and the biosphere to change. So whether it be the Earth's temperature staying within the narrow range needed to maintain life over billions of years, while the Sun's output has risen by 30% and the continents have moved across the face of the planet, or the ability of life to adapt to the catastrophe of a large bolide impact, it is helpful to consider the total response to these challenges. The long-term changes of the composition of the atmosphere and the waxing and waning of the ice ages are other examples of how we must consider all the issues if we are to form a sensible picture.

None of this makes climate change any easier to understand, but it may help to maintain a balanced attitude to what has happened; what is happening now; and what could happen in the future. Equally important, by taking a suitably wide view we may reinforce a sense of wonderment for the immensity and complexity of the Earth's climate.

QUESTIONS

1   The term 'consensus' used to describe the agreement reached by the IPCC on the impact of human activities on the climate has been described as being anything from a courageous scientific compromise through an uneasy

truce to an unfortunate if not politically inspired fudge. On your reading of the report what do you think is the best way to describe the conclusion and why?

2   Do you think there is any way in which political parties in both developed and developing countries will reconcile the pressures for economic growth with the demands to cut back emissions of greenhouse gases if the only solution is to reduce growth and employment before there is incontrovertible evidence of climate damage?

3   Consider the arguments for and against taking specific action to modify the climate on a regional or global scale (e.g. programmes of afforestation, or diversion of ocean currents). What issues would need to be resolved before such large-scale activities were put in train?

4   Identify possible low cost ways of minimising the impact of global warming in urban areas, with particular regard to cutting the demand for air-conditioning and reducing air pollution. What are the economic and social barriers to introducing these options which could reduce the amount of solar energy absorbed and cut down vehicle emissions?

5   If, once every million years (see Fig. 8.11) or so, a bolide impact with the Earth causes major climatic damage, how much money do you think we should spend to protect ourselves against such an event, and what would you recommend we spend the money on?

FURTHER READING

A complete reference list is available at the end of the book but the following is a selection of the best books or articles to follow up particular topics within this chapter. Full details of each reference are to be found in the Bibliography.

Houghton (1997). A penetrating and committed analysis of the issues surrounding global warming and the options for reducing its impact, written by the co-chairman of Scientific Assessment Working Group of the IPCC, which is required reading for anyone who wants an accessible presentation of the arguments for action without going through all the IPCC reports.

Lovelock (1988). The best source for an up-to-date analysis of the basic content of the Gaia hypothesis of the stability of the Earth's biosphere.

Mintzer (1992). A series of papers by leading authorities in their areas on the various economic and social challenges which will have to be addressed if we are to come to terms with the predicted levels of global warming in the next century.

BIBLIOGRAPHY

Alley, R. B., et al. (1993). Abrupt increase in Greenland snow accumulation at the end of the Younger Dryas event. *Nature*, **362**, 527–9.

Baillie, M. G. L. (1995). *A Slice Through Time: Dendrochronology and Precision Dating*. Batsford, London, UK.

Barber, D. C., et al. (1999). Forcing of the cold event of 8,200 years ago by catastrophic drainage of Laurentide lakes. *Nature*, **400**, 344–8.

Barnston, A. G. (1995). Our improving capability in ENSO forecasting. *Weather*, **50**, 419–30.

Barry, R. G. & Chorley, R. J. (1992). *Atmosphere, Weather & Climate. Sixth Edition*. Methuen, London, UK.

Benton, M. J. (1995). Diversification and extinction in the history of life. *Science*, **268**, 52–8.

Bigg, G. R. (1996). *The Oceans and Climate*. Cambridge University Press, Cambridge, UK.

Bonan, G. B., Pollard, D. & Thompson, S. L. (1992). Effects of boreal forest vegetation on global climate. *Nature*, **359**, 716–18.

Bradley, R. S. & Jones, P. D. (eds) (1995). *Climate Since AD 1500*. Routledge, London, UK.

Briffa, K. R., et al. (1990). A 1400-year tree-ring record of summer temperatures in Fennoscandia. *Nature*, **346**, 434–9.

Briffa, K. R., et al. (1995). Unusual twentieth-century summer warmth in a 1000-year temperature record from Siberia. *Nature*, **376**, 156–9.

Broecker, W. S. (1994). Massive iceberg discharges as triggers for global climate change. *Nature*, **372**, 421–5.

Broecker, W. S. (1995a). Chaotic climate. *Scientific American*, **267**, No. 11, 44–50.

Broecker, W. S. (1995b). Cooling the tropics. *Nature*, **376**, 212–13.

Brown, G. C., Hawkesworth, C. J. & Wilson, R. C. L. (eds) (1992). *Understanding the Earth: A New Synthesis*. Cambridge University Press, Cambridge, UK.

Bryant, E. (1997). *Climate Process and Change*. Cambridge University Press, Cambridge, UK.

Burroughs, W. J. (1978). On running means and meteorological cycles. *Weather*, **33**, 101–9.

Burroughs, W. J. (1991). *Watching the World's Weather*. Cambridge University Press, Cambridge, UK.

Burroughs, W. J. (1994). *Weather Cycles: Real or Imaginary?* Cambridge University Press, Cambridge, UK.

Burroughs, W. J. (1997). *Does the Weather Really Matter?* Cambridge University Press, Cambridge, UK.

Burroughs, W. J. (1999). *The Climate Revealed.* Mitchell-Beazley, London, UK.

Burroughs, W. J. & Lynagh, N. (1999). *Maritime Weather and Climate.* Witherby, London, UK.

Cess, R. D., et al. (1991). Interpretation of snow-climate feedback as produced by 17 general circulation models. *Science*, **253**, 888–92.

Cess, R. D., et al. (1995). Absorption of solar radiation by clouds: observations versus models. *Science*, **267**, 496–9.

Chahine, M. T. (1992). The hydrological cycle and its influence on the climate. *Nature*, **359** 373–9.

CLIMAP (1976). The surface of the ice-age Earth. *Science*, **191**, 1131–7.

Craddock, J. M. (1968). *Statistics in the Computer Age.* English University Press, London, UK.

Currie, R. G. (1988). Lunar tides and the wealth of nations. *New Scientist*, **120**, 5 November, 52–5.

Dansgaard, W., Johnsen, S. J., Reeh, N., Gunderstrup, N., Clausen, H. B. & Hammer, C. U. (1975). Climatic changes, Norsemen and modern man. *Nature*, **255**, 24–8.

Dansgaard, W., et al. (1993). Evidence of general instability of past climate from a 250-kyr ice-core record. *Nature*, **364**, 218–20.

Dawson, A. G. (1992). *Ice Age Earth: Late Quaternary Geology and Climate.* Routledge, London, UK.

Doherty, R. M., Hulme, M. & Jones, C. G. (1999). A gridded reconstruction of land and ocean precipitation for the extended tropics from 1974–1994. *International Journal of Climatology*, **19**, 119–42.

Elsner, J. B. & Tsonis, A. A. (1991). Do bidecadal oscillations exist in the global temperature record? *Nature*, **353**, 551–3

Engelen, A. F. V. van & Nellestijn, J. W. (1995). Monthly, seasonal, and annual means of the air temperature in tenths of centigrade in De Bilt, Netherlands, 1706–1995. KNMI.

Folland, C. K. and Parker, D. E. (1995). Correction of instrumental biases in historical sea surface temperature data. *Quarterly Journal Royal Meteorological Society*, **121**, 319–67.

Frakes, L. A., Francis, J. E. & Syktus, J. I. (1992). *Climate Modes of the Phanerozoic.* Cambridge University Press, Cambridge, UK.

Fritts, H. C. (1976). *Tree Rings and Climate.* Academic Press, London, UK.

Giovanelli, R. (1984). *Secrets of the Sun.* Cambridge University Press, Cambridge, UK.

Gray, W. M. (1990). Strong association between West African rainfall and US landfall of intense hurricanes. *Science*, **249**, 1251–6.

Greenland Ice Core Project (GRIP) Members (1993). Climate instability during the last interglacial period recorded in the GRIP ice core. *Nature*, **364**, 203–7.

Grootes, P. M., Stuiver, M., White, J. W. C., Johnsen, S. & Jouzel, J. (1993). Comparison of oxygen isotope records from the GISP 2 and Greenland ice cores. *Nature*, **366**, 552–4.

Grove, J. M. (1988). *The Little Ice Age.* Methuen, London, UK.

Gurney, R. J., Foster, J. L. & Parkinson, C. L. (eds) (1993). *Atlas of Satellite Observations Related to Global Change.* Cambridge University Press, Cambridge, UK.

Haigh, J. D. (1996). The impact of solar variability on climate. *Science*, **272**, 981–4.

Hammer, C. U., Clausen, H. B. & Dansgaard, W. (1980). Greenland ice sheet evidence of post-glacial volcanism and its climatic impact. *Nature*, **288**, 230–5.

Hansen, J. E., Wilson, H., Sato, M., Ruedy, R., Shah, K. P. & Hansen, E. (1995). Satellite and surface temperature data at odds? *Climate Change*, **30**, 103–17.

Harries, J. E. (1990). *Earthwatch*. Ellis Horwood, Chichester, UK.

Hostetler, S. W. & Mix, A. C. (1999). Reassessment of ice-age cooling of the tropical ocean and atmosphere. *Nature*, **399**, 673–6.

Houghton, J. T. (1986). *The Physics of Atmospheres. Second Edition*. Cambridge University Press, Cambridge, UK.

Houghton, J. (1997). *Global Warming: The Complete Briefing*. Cambridge University Press, Cambridge, UK.

Hulme, M. & Barrow, E. (eds) (1997). *Climates of the British Isles: Present, Past and Future*. Routledge, London, UK.

Hulme, M. & Jones, P. D. (1991). Temperatures and windiness over the United Kingdom during the winters of 1988/89 and 1989/90 compared with previous years. *Weather*, **46**, 126–36.

Hurrell, J. W. (1995). Decadal trends in the North Atlantic Oscillation: Regional temperatures and precipitation. *Science*, **269**, 676–9.

Hurrell, J. W. (1996). Influence of variations in extratropical wintertime tele-connections on Northern Hemisphere temperature. *Geophysical Research Letters*, **23**, 665–8.

Imbrie, J. & Imbrie, J. Z. (1979). *Ices Ages: Solving the Mystery*. Macmillan, London, UK.

Imbrie, J. & Imbrie, J. Z. (1980). Modelling the climatic response of orbital variations. *Science*, **207**, 943–53.

Imbrie, J., et al. (1992). On the structure and origin of major glaciation cycles. 1. Linear responses to Milankovitch forcing. *Paleoceanography*, **7**, 701–38.

Imbrie, J., et al. (1993). On the structure and origin of major glaciation cycles. 2. The 100,000-year cycle. *Paleoceanography*, **8**, 699–735.

IPCC (1990). *Climate Change: The IPCC Scientific Assessment*, J. T. Houghton, G. J. Jenkins & G. G. Ephraums (eds). Cambridge University Press, Cambridge, UK, 365 p.

IPCC (1992). *Climate Change 1992: The Supplementary Report to IPCC Scientific Assessment*, J. T. Houghton, B. A. Callander & S. K. Varney (eds). Cambridge University Press, Cambridge, UK.

IPCC (1994). *Climate Change 1994: Radiative Forcing of Climate and an Evaluation of the IPCC IS92 Emission Scenarios*, J. T. Houghton, L. G. Meira Filho, J. Bruce, H.-S. Lee, B. A. Callander, E. Haites, N. Harris & K. Maskell (eds). Cambridge University Press, Cambridge, UK.

IPCC (1995). *Climate Change 1995: The Science of Climate Change*, J. T. Houghton, L. G. Meira Filho, B. A. Callander, N. Harris, A. Kattenberg & K. Maskell (eds). Cambridge University Press, Cambridge, UK.

International Federation of Red Cross and Red Crescent Societies (1999). *World Disasters Report*.

Jablonski, D. (1997). Progress at the K-T boundary. *Nature*, **387**, 354–5.

Jolliffe, I. T. (1986). *Principal Component Analysis*. Springer-Verlag, New York, USA.

Jones, P. D., New, M., Parker, D. E., Martin, S. & Rigor, I. G. (1999). *Review of Geophysics*, **37**, 173–99.

Karl, T. R., et al. (1995). Critical issues for long-term climate monitoring. *Climatic Change*, **31**, 185–221.

Karl, T. R., Knight, R. W., Easterling, D. R. & Quayle, R. G. (1996). Indices of climate change for the United States. *Bulletin of the American Meteorological Society*, **77**, 279–92.

Kates, R. W., Ausubel, J. H. & Berberain, M. (eds) (1985). *Climate Impact Assessment: Studies of the Interaction of Climate and Society*. Wiley, Chichester, UK.

Kendall, M. (1976). *Time Series*. Charles Griffin, London, UK.

Kerr, R. A. (1995). Darker clouds promise brighter future for climate models. *Science*, **267**, 454.

Kripalani, R.H. & Kulkarni, A. (1997). Climatic impact of the El Niño/La Niña on the Indian monsoon: a new perspective. *Weather*, **52**, 39–46.

Krishna Kumar, K., Soman, M. K. & Rupa Kuma, K. (1995). Seasonal forecasting of Indian summer monsoon rainfall: a review. *Weather*, **50**, 449–67.

Laird, K. R., Fritz, S. C., Maasch, K. A. & Cumming B. F. (1996). Greater drought intensity and frequency before AD 1200 in the Northern Great Plains, USA. *Nature*, **384**, 552–4.

LaMarche, V. C. Jr. (1974). Paleoclimatic inferences from long tree-ring records. *Science*, **183**, 1043–8.

Lamb, H. H. (1972). *Climate: Present, Past and Future*. Volume 1. Methuen, London, UK.

Lamb, H. H. (1977). *Climate: Present, Past and Future*. Volume 2. Methuen, London, UK.

Lamb, H. H. (1995). *Climate, History and the Modern World* (2nd edn). Routledge, London, UK.

Landsea, C. W. (1993). A climatology of intense (or major) Atlantic hurricanes. *Monthly Weather Review*, **121**, 1703–13.

Landsea, C. W., Gray, W. M., Mielke, P. W. Jr. & Berry, J. K. (1994). Seasonal forecasting of Atlantic hurricane activity. *Weather*, **49**, 273–84.

Le Roy Ladurie, E. & Baulant, M. (1980). Grape harvests from the fifteenth through the nineteenth centuries. *Journal of Interdisciplinary History*, **10**, 839–49.

Li, X., Maring, H., Savoie, D., Voss, K. & Prospero, J. M. (1996). Dominance of mineral dust in aerosol light scattering in the North Atlantic trade winds. *Nature*, **380**, 416–19.

Lovelock, J. E. (1988). *The Ages of Gaia*. Oxford University Press, Oxford, UK.

Manley, G. (1974). Central England Temperatures: monthly means 1659 to 1973. *Quarterly Journal Royal Meteorological Society*, **100**, 389–405.

Markson, R. (1978). Solar modification of atmospheric electrification and possible implications for the Sun–weather relationship. *Nature*, **244**, 197–200.

Martinson, D. G., et al. (1987). Age dating and the orbital theory of the Ice Age: development of a high resolution 0 to 300,000-year chronostratigraphy. *Quaternary Research*, **17**, 1–30.

McIlveen, R. (1992). *Fundamentals of Weather and Climate*. Chapman and Hall, London, UK.

Mintzer, I. M. (ed) (1992). *Confronting Climate Change: Risks, Implications and Responses*. Cambridge University Press, Cambridge, UK.

Mitchell, J. F. B., Johns, T. C., Gregory, J. M. & Tett, S. F. B. (1995). Climate response to increasing levels of greenhouse gases and sulphate aerosols. *Nature*, **376**, 501–4.

Mitchell, J. M. (1990) Climatic variability: past, present & future. *Climatic Change*, **16**, 231–46.

Mitchell, J. M., Stockton, C. W. & Meko, D. M. (1979). Evidence of a 22-year rhythm of drought in the Western United States related to the Hale Solar Cycle since the 17th century. In *Solar-Terrestrial Influences on Weather and Climate*, B. M. McCormac & T. A. Seliga (eds). D. Reidel Publishing Co., Dordrecht, Nl.

Musk, L. F. (1988). *Weather Systems*. Cambridge University Press, Cambridge, UK.

Palmer, T. (1993). A nonlinear dynamical perspective on climate change. *Weather*, **48**, 314–25.

Parker, D. E., Legg, T. P. & Folland, C. K. (1992). A new daily Central England temperature series, 1772–1991. *International Journal of Climatology*, **12**, 317–42.

Parry, M. & Duncan, R. (eds) (1995). *The Economic Implications of Climate Change in Britain*. Earthscan, London, UK.

Pecker, J.-C. & Runcorn, S. K. (eds) (1990). *The Earth's Climate and Variability of the Sun over Recent Millennia: Geophysical, Astronomical and Archaeological Aspects*. The Royal Society, London, UK (printed by CUP).

Petit, J. R., et al. (1999). Climate and atmospheric history of the past 420,000 years from the Vostok ice core, Antarctica. *Nature*, **399**, 429–36.

Pfister, C. (1995). Monthly temperature and precipitation in central Europe 1525–1979: quantifying documentary evidence on weather and its effects. In *Climate Since AD 1500*, Chapter 6, R. S. Bradley & P. D. Jones (eds). Routledge, London, UK. (Data on temperature and precipitation indices available on disk from the National Geophysical Data Center, Boulder, CO 80309, USA.)

Philander, S. G. (1990). *El Niño, La Niña and the Southern Oscillation*. Academic Press, San Diego, USA.

Pielke, R. A. & Landsea, C. W. (1998). Normalized hurricane damage in the United States: 1925–95. *Weather and Forecasting*, **13**, 621–31.

Pimental, D., et al. (1995). Environmental and economic costs of soil erosion and conservation benefits. *Science*, **267**, 1117–23.

Rampino, M. R. & Self, S. (1992). Volcanic winter and accelerated glaciation following the Toba super-eruption. *Nature*, **359**, 50–2.

Raup, D. M. (1991). *Extinction: Bad Genes or Bad Luck?* W. W. Norton, New York, USA.

Robin, G. de Q. (1983). *The Climatic Record in Polar Ice Sheets*. Cambridge University Press, Cambridge, UK.

Rosenzweig, C. & Parry, M. L. (1994). Potential impact of climate change on world food supply. *Nature*, **367**, 133–8.

Shoemaker, E. M. (1983). Asteroid and Comet Bombardment of the Earth. *Annual Review of Earth and Planetary Science*, **11**, 461–94.

Spencer, R. W. & Christy, J. R. (1990). Precise monitoring of global temperature trends from satellites. *Science*, **247**, 1558–62.

Stommel, H. & Stommel, E. (1979). The year without a summer. *Scientific American*, **240**, June, 134–40.

Stothers, R. B. (1984). Mystery cloud of AD 536. *Nature*, **307**, 344–5.

Stouffer, R. J., Manabe, S. & Vinnikov, K. Ya. (1994). Model assessment of the role of natural variability in recent global warming. *Nature*, **367**, 634–6.

Strzepek, K. M. & Smith, J. B. (eds) (1995). *As Climate Changes: International Impacts and Implications*. Cambridge University Press, Cambridge, UK.

Taylor, K. C., et al. (1993). The 'flickering switch' of late Pleistocene climate change. *Nature*, **361**, 432–6.

Tengen, I., Lacis, A. A. & Fung, I. (1996). The influence on climate forcing of mineral aerosols from disturbed soils. *Nature*, **380**, 419–22.

Tett, S. F. B., Stott, P. A., Allen, M. R., Ingram, W. J. and Mitchell, J. F. B. (1999). Causes of twentieth century climate change. *Nature*, **399**, 569–72.

Thomas, D. S. G. & Middleton, N. J. (1994). *Desertification: Exploding the Myth*. Wiley, Chichester, UK.

Toumi, R., Bekki, S. & Law, K. (1995). Indirect influence of ozone depletion on climate forcing by clouds. *Nature*, **372**, 348–51.

Trenberth, K. E. (ed.) (1992). *Climate System Modeling*. Cambridge University Press, Cambridge, UK.

Tucker, C. J., Dregne, H. E. & Newcomb, W. W. (1991). Expansion and contraction of the Sahara Desert from 1980 to 1990. *Science*, **253**, 299–301.

Van Andel, T. H. (1994). *New Views on an Old Planet: A History of Global Change*. Cambridge University Press, Cambridge, UK.

Van den Dool, H. M., Kriijnen, H. J. & Schuurmans, C. J. E. (1978). Average winter temperatures at De Bilt (Netherlands): 1634–1977. *Climatic Change*, **1**, 319–30.

Weaver, A. J. & Hughes, T. M. C. (1994). Rapid interglacial climate fluctuations driven by North Atlantic ocean circulation. *Nature*, **367**, 447–50.

Weaver, A. J., Sarachik, E. S. & Marotzke, J. (1991). Freshwater flux forcing of decadal and interdecadal oceanic variability. *Nature*, **353**, 836–8.

Whitlock, C. & Bartlein, P. J. (1997). Vegetation and climate change in northwest America during the last 125 kyr. *Nature*, **388**, 57–61.

Wigley, T. M. L., Ingram, M. J. & Farmer, G. (1981). *Climate and History: Studies in Past Climates and their Impact on Man.* Cambridge University Press, Cambridge, UK.

Wigley, T. M. L., Lough, J. M. & Jones, P. D. (1984). Spatial patterns of precipitation in England and Wales and a revised homogenous England and Wales precipitation series. *Journal of Climatology*, **4**, 1–25.

Willson, R. C. & Hudson, H. S. (1991). The Sun's luminosity over a complete cycle. *Nature*, **351**, 42–4.

Wiscombe, W. J. (1995). An absorbing mystery. *Nature*, **376**, 466–7.

GLOSSARY

Absolute temperature (K)   A temperature scale based on the thermodynamic principle that the lowest possible temperature is absolute zero (0 K) and the ice point is 273.16 K measured in degrees Kelvin (K) which are of the same magnitude as degrees Celsius (°C).

Absorption   The process by which incident radiation is taken into a body and retained without reflection or transmission, thereby increasing the internal or kinetic energy of the molecules or atoms composing the absorbing medium.

Aerosols   Particles, other than water or ice, suspended in the atmosphere. They range in radius from one hundredth to one ten-millionth of a centimetre – or $10^2 - 10^{-3}$ micrometres (μm). Aerosols are important as nuclei for the condensation of water droplets and ice crystals, and as participants in various atmospheric chemical reactions. Aerosols resulting from volcanic eruptions can lead to a cooling at the Earth's surface.

Albedo   The proportion of the radiation falling upon a non-luminous body which it diffusely reflects.

Angiosperms   Flowering plants; first arrived in the Cretaceous.

Biosphere   The system of earth and its atmosphere that supports life. In the global carbon cycle, the biosphere serves as a sink (reservoir): carbon is stored and preserved in living organisms (plants and animals) and life-derived organic matter (litter, detritus). The biosphere controls the magnitude of the fluxes of several greenhouse gases, including CO_2 and methane, between the atmosphere, oceans and land. The terrestrial biosphere includes living biota (plants and animals) and the litter and soil organic matter on land. The marine biosphere includes the flora, fauna and detritus in the oceans.

Black body radiation   The radiation that is emitted by a surface which absorbs all incident radiation at all wavelengths. The wavelength dependence of this radiation is defined by the temperature of the surface.

Blocking   A phenomenon, most often association with stationary high pressure systems in the mid-latitudes of the northern hemisphere, which produces periods of abnormal weather.

Bolide   Any solid object from outer space; a term usually restricted to objects larger than common meteorites.

Carbonate sediments   Sediments composed of the calcium carbonate ($CaCO_3$) minerals aragonite and calcite or the calcium–magnesium carbonate ($CaMg(CO_3)_2$) dolomite.

Chlorofluorocarbons (CFCs)   A family of inert, non-toxic and easily liquefied chemicals, implicated in two major environmental problems: ozone-layer depletion and global warming. CFCs are used as coolants in refrigerators and air-conditioners, as propellants in aerosol cans, as solvents and as blowing-agents that inflate flexible foams. It takes about fifteen years for a CFC molecule to drift into the upper atmosphere, where it can last 100 years or more, destroying an estimated 10,000 ozone molecules over that time.

Climate   The long-term statistical average of weather conditions. Global climate represents the long-term behaviour of such parameters as temperature, air pressure, precipitation, soil moisture, runoff, cloudiness, storm activity, winds and ocean currents, integrated over the full surface of the globe. Regional climate, analogously, are the long-term averages for geographically limited domains on the Earth's surface.

Coccolithophorida   A group of plankton that construct a calcareous shell of round platelets known as coccoliths, which may accumulate on the ocean floor as sediment, ultimately contributing to limestone, such as chalk (see also phytoplankton).

Cold front   The boundary line between advancing cold air and a mass of warm air under which the cold air pushes like a wedge.

Continental drift   The lateral movement of continents as a result of sea-floor spreading (see also Plate tectonics).

Convection   A type of heat transfer which occurs in a fluid by the vertical motion of large volumes of the heated material by differential heating (at the bottom of the atmosphere) thus creating, locally, a less dense, more buoyant fluid.

Coriolis force   The term used to explain the fact that a moving object detached from the rotating Earth appears to an observer on Earth to be deflected by a force acting at right angles to the direction of motion. Deflection of moving objects in the Northern Hemisphere is to the right of the path of motion. Deflection in the Southern Hemisphere is to the left of the path of motion.

Cretaceous (65–146 Ma)   An era in which mean global temperatures may have been 10 °C warmer than they are today, and deep water temperatures may have been 18 °C warmer. The last period of the Mesozoic era, the Cretaceous was marked by the rapid rise and spread of deciduous trees, by shallow seas submerging most of the Earth's present land surface, and by dinosaurs – which became extinct after this period.

Cryosphere   The portion of the climate system consisting of the world's ice masses, sea ice, glaciers and snow deposits. Snow cover on land is largely seasonal and related to atmospheric circulation. Glaciers and ice sheets are tied to global water cycles and variations of sea level, and change over periods from hundreds to millions of years. The ice sheets of Greenland and the Antarctic contain 80% of the existing freshwater on the globe, thereby acting as a long-term reservoir in the hydrological cycle.

Deforestation   Loss of forest. At least eleven million hectares of tropical forest are lost every year. Although the causes vary by region, one estimate indicates that slash-and-burn agriculture and scavenging for wood-fuel, often in the wake of commercial road-building, accounts worldwide for 40–50% of deforestation. Grazing accounts for 10%, commercial agriculture for 10–20%, forestry and plantations for 5–10%, and forest fires for 1–15%.

Dendrochronology   The dating of past events and variations in the environment and climate by studying the annual growth rates of trees.

Dendroclimatology   The science of reconstructing past climates from the information stored in tree trunks as the annual radial increments of growth. Wide rings signify favourable growing conditions, absence of disease and pests, and favourable climatic conditions. Narrow rings indicate unfavourable growing conditions or climate. Tree rings record responses to a wider range of climatic variables.

Depression   A part of the atmosphere where the surface pressure is lower than in surrounding parts – often called a 'low'.

Ecliptic   The great circle in which the plane containing the centres of the Earth and the Sun cuts the celestial sphere.

Eemian (117,000–130,000 y BP)   The interglacial optimum period prior to the current one when it was warmer than today. This warmth may have been a consequence of the Earth's orbit giving markedly more radiation during the Northern hemisphere summer.

Electromagnetic radiation   The emission and propagation of electromagnetic energy from a source in the form of electric and magnetic fields, which need no medium to support them and which travel through a vacuum at the velocity of light. This radiation encompasses the entire frequency range from γ-rays to radiowaves.

El Niño Southern Oscillation (ENSO)   A quasi-periodic occurrence when large-scale abnormal pressure and sea-surface temperature patterns become established across the tropical Pacific every few years.

Emissivity   The ratio of the emissive power of a surface at a given temperature to that of a black body at the same temperature and the same surroundings.

Eocene (37–55 Ma)   An epoch, part of the Tertiary era. Like the Cretaceous era, the Eocene epoch was significantly warmer than the present day. Seas expanded far beyond their present boundaries; palm trees grew where London and southern Alaska are today.

Eustasy   The worldwide global changes in sea level caused by changes in water volume due to the formation and melting of ice sheets, changes in the temperature of the water, or changes in the volume of ocean basins induced by changing volumes of ocean ridges.

Eustatic   A global change in sea level.

Evapotranspiration   The process of water vapour transfer from vegetated land surfaces into the atmosphere; an essential part of the global hydrologic cycle. Evapotranspiration includes *evaporation* (the change of liquid water, from bodies of water and wet soil, into water vapour) and *transpiration* (in which water is drawn from the soil into plant roots, transported through the plant, and then evaporated from leaves and other plant surfaces into the air).

Feedback mechanism   A process of system dynamics in which a system reacts to amplify or suppress the effect of a force which is acting upon it. For example, in the climate system, warmer temperatures may melt snow and ice cover, revealing the darker land surface underneath. The darker surface absorbs more solar energy, causing further temperature increases, thus melting even more snow and ice cover and so on. This is positive feedback, in which warming reinforces itself. In negative feedback, a force ultimately reduces its own effect. For example, when the Earth's surface grows warmer, more water evaporates, forming more clouds. If the clouds which form are extensive and widely distributed, covering large areas of the surface, they will tend to reflect more solar radiation back into space than the dark ground underneath would, cooling the Earth's surface and reducing the impact of warmer temperatures.

Foraminifera (foraminifers, or forams for short)   A group of microscopic marine organisms belonging to the Protozoa; they are planktonic or benthonic and are grazers and predators. Their calcareous shells provide a major part of the palaeo-ceanographic and palaeoclimatic record.

Fourier transform spectral analysis   The mathematical determination of the amplitude of the harmonic components of a time series and the presentation of these in the form of a power spectrum (see Power spectrum).

Gaia hypothesis   This hypothesis holds that living organisms on Earth (including micro-organisms) actively regulate atmospheric composition and climate, helping provide climate stability in the face of challenges like the increasing luminosity of the sun, or increasing anthropogenic greenhouse gas emissions.

General Circulation Models (GCMs)   A computational model or representation of the Earth's climate used to forecast changes in climate or weather.

Glacial epochs   Periods during the history of the Earth when there were larger ice sheets (continental-size) and mountain glaciers than today. The most recent glacial epoch, the Pleistocene, has encompassed much of the last 2.5 My. In overall occurrence, all the glacial epochs that have ever occurred occupy only 5–10% of all geologic time. During major glacial epochs, which seem to recur at intervals of 200 to 250 My, great ice sheets form in the high latitudes and spread out to cover as much as 40% of the Earth's land surface.

Glacial rebound (See Isostasy.)

Greenhouse effect   An atmospheric process in which the concentration of atmospheric trace gases (greenhouse gases) affects the amount of radiation that escapes directly into space from the lower atmosphere. Short-wave solar radiation can pass through the clear atmosphere relatively unimpeded. But long-wave terrestrial radiation, emitted by the warm surface of the Earth, is partially absorbed and then re-emitted by certain trace gases.

Greenhouse gases   The trace gases which contribute to the greenhouse effect. The main greenhouse gases are not the major constituents of the atmosphere – nitrogen and oxygen – but water vapour (the biggest contributor), carbon dioxide, methane, nitrous oxide and (in recent years), chlorofluorocarbons. Increases in concentrations of the latter four gases have been linked to human activity.

Hadley cell   The basic vertical circulation pattern in the tropics where moist warm air rises near the equator and spreads out north and south and descends at around 20–30° N and S.

Hale cycle   The twenty-two-year cycle in solar activity which is a combination of the eleven-year cycle in sunspot number and the reversal of the magnetic polarity of adjacent pairs of sunspots between alternate cycles, which may be a cause of the twenty-year cycle which is detected in many climatic records.

Half life   Time in which half of the atoms of a given quantity of radioactive nuclide undergo at least one disintegration.

Holocene   The relatively warm epoch, which started around 10,000 years ago and runs up to present time. It is marked by several short-lived particularly warm periods, the most significant of which, from 6,200–5,300 y BP, is called the Holocene optimum.

Hurricane   The name given primarily to tropical cyclones in the West Indies and Gulf of Mexico.

Insolation (from INcoming SOLar radiATION)   The solar radiation received at any particular area of the Earth's surface, which varies from region to region depending on latitude and weather.

Interglacial   Warmer periods during **Glacial epochs** when the major ice sheets recede to higher latitudes.

Interstadial   A relatively warmer stage within a glacial phase during which the ice advance is temporarily halted.

Intertropical Convergence Zone (ITCZ)   A narrow low-latitude zone in which air masses originating in the northern and southern hemispheres converge and generally produce cloudy, showery weather. Over the Atlantic and Pacific it is the boundary between the north-east and south-east trade winds. The mean position is somewhat north of the equator but over the continents the range of movement throughout the year is considerable.

Isostasy   The process whereby areas of the crust tend to float in conditions of near equilibrium on the plastic mantle, and where ice sheets have melted this process leads to a slow rise in the crust as it returns to equilibrium (glacial rebound).

Isotopes   Atoms of a single element (with the same number of protons) that have different masses (because they have a different number of neutrons). Isotopes are labelled with the atomic mass preceding the symbol of the element: ^{18}O, for example, denotes an oxygen isotope with an atomic mass of 18, instead of the mass of 16 which the most abundant form of oxygen has (as ^{16}O). Some isotopes have characteristics making them useful for analysing chemical history; for example, they release electrons or other sub-atomic particles, allowing their presence to be detected, and they decay (gradually changing into another element) at a steady, measurable speed. Carbon-14 (^{14}C) decays into ^{14}N, with a half life of approximately 5,700 years. When detected in samples of sediment or ice, isotopes can be analysed to compile records of past climate characteristics.

Jet stream   Strong winds in the upper troposphere whose course is related to the major weather systems in the lower atmosphere and which tend to define the movement of these systems.

Kelvin waves   Gravity-inertia waves which occur in both the atmosphere and the oceans, where either the effect of the Coriolis Force is negligible (i.e. close to the equator) or where this force is balanced by the pressure gradient. The most important examples are in the equatorial stratosphere and in the thermocline of the equatorial Atlantic and Pacific close to the equator (in both cases the waves propagate eastwards relative to the Earth).

Last Glacial Maximum (18,000 y BP)   The last prolonged period of Ice Age cold climate before the present day.

Lithification   The process of conversion of an unconsolidated sediment to a solid rock.

Lithosphere   The outer rigid shell of the Earth.

Little Ice Age (AD 1550–AD 1850)   A period marked by more frequent cold episodes in Europe, North America and Asia, during which mountain glaciers, especially in the Alps, Norway, Iceland and Alaska expanded substantially.

Magma   Naturally occurring molten or partially molten rock formed within the Earth from which igneous rocks crystallize.

Maunder minimum   A period during the seventeenth century when the level of solar activity, as reflected by the number of sunspots, was much lower than in subsequent centuries.

Mean sea level (MSL)   The average height of the sea surface, based on hourly observation of the tide height on the open coast, or in adjacent waters that have free access to the sea. In the United States, MSL is defined as the average height of the sea surface for all stages of the tide over a nineteen-year period.

Micrometre (μm) 10^{-6} m.

Monsoon   A seasonal reversal of wind which in the summer season blows onshore, bringing with it heavy rains, and in winter blows offshore – it is of greatest meteorological importance in southern Asia. The word is believed to be derived from the Arabic word '*mausin*', meaning a season.

Non-linearity   The lack of direct proportionality of the input and output of a physical system.

North Atlantic Oscillation (NAO)   An index of the circulation in the North Atlantic which is measured in terms of the difference in pressure between the Azores and Iceland. In winter this index tends to switch between a strong westerly flow with pressure low to the north and high in the south and a weaker opposite pattern: the former tends to produce above normal temperatures over much of the northern hemisphere, the latter the reverse.

Nutation   Oscillation of the Earth's pole about the mean position. It has a period of about 19,000 years and is superimposed on the precessional movement.

Obliquity of the ecliptic   The angle at which celestial equator intersects the ecliptic. At present this angle is slowly decreasing by 0.47 arc seconds a year, owing to precession and nutation. It varies between 21° 53′ and 24° 18′.

Ozone   A molecule made up of three atoms of oxygen (O_3). In the stratosphere, it occurs naturally and provides a protective layer shielding the earth from ultraviolet radiation and subsequent harmful health effects on humans and the environment. In the troposphere, it is a major component of photochemical smog.

Pelagic ooze   A deep ocean sediment formed from the hard parts of pelagic organisms and very fine suspended sediment.

Photosynthesis   The process by which plants convert carbon dioxide (CO_2) and water (H_2O) of the air into carbohydrates by exposure to light.

Phytoplankton   Microscopic marine organisms (mostly algae and diatoms) which are responsible for most of the photosynthetic activity in the oceans.

Plate tectonics   The interpretation of the Earth's structures and processes (including oceanic trenches, mid-ocean ridges, mountain building, earthquake zones and volcanic belts) in terms of the movement of large plates of the lithosphere acting as rigid slabs floating on a viscous mantle.

Pleistocene (10,000–1,600,000 y BP)   The geological period, which together with the Holocene makes up the Quaternary. This epoch was characterised by numerous (at least seventeen) worldwide changes of climate, cycling between glacial (cool) and interglacial (warmer) periods, with periodicities of 100,000, 41,000 and 23,000 years.

Pliocene (1.6–5.3 Ma)   A warm epoch with only limited glaciation. The mid-Pliocene was the last time that comparable temperatures existed to those predicted by General Circulation Models to take place within 300 years. Precipitation was greater than at present, including in the arid regions of Middle East, Asia and Northern Africa, where temperatures were lower than at present in summer.

Power spectrum   The presentation of the square of the amplitudes of the harmonics of a time series as a function of the frequency of the harmonics.

Precession of the equinoxes   The westward motion of 50.27 arc seconds per year of the equinoxes, caused mainly by the attraction of the Sun and the Moon on the equatorial bulge of the Earth. The equinoxes thus make one complete revolution of the ecliptic in 25,800 years and the Earth's pole turns in a small circle of radius 23° 27' about the pole of the ecliptic.

Proxy data   Any source of information which contains indirect evidence of past changes in the weather (e.g. tree rings, ice cores and ocean sediments).

Quasi-biennial oscillation (QBO)   The alternation of easterly and westerly winds in the equatorial stratosphere with an interval between successive corresponding maxima of twenty to thirty-six months. Each new regime starts above 30 km and propagates downwards at about 1 km a month.

Quaternary   The geological period covering the last 1.6 million years or so, which includes the Pleistocene and the Holocene.

Radiative forcing   A change imposed upon the climate system which modifies the radiative balance of that system. The causes of such a change may include changes in the sun, clouds, ice, greenhouse gases, volcanic activity and other agents. Convention lumps all these together as agents of radiative forcing. Radiative forcing is often specified as the net change in energy flux at the tropopause (watts per square meter [$W\,m^{-2}$]). Many climate models seek to quantify changes in Earth's temperature, rainfall and sea level in terms of a specified change in radiative forcing.

Radiatively active trace gases   Gases, present in small quantities in the atmosphere, that absorb incoming solar radiation or outgoing infrared radiation, thus affecting the vertical temperature profile of the atmosphere. These gases include water vapour, carbon dioxide, methane, nitrous oxide, chlorofluorocarbons and ozone.

Radiolaria   Marine pelagic micro-organisms with a siliceous skeleton.

Radiometer   An instrument which makes quantitative measurements of the amount of electromagnetic radiation falling on it in a specified wavelength interval.

Radiometric date   The age of a rock in years determined by the relative proportions of radioactive isotopes and their decay products present within the rock.

Radio-sonde   A free balloon carrying instruments which transmit measurements of temperature, pressure and humidity to ground by radiotelegraphy as it rises through the atmosphere.

Rossby wave   In the atmosphere a wave in the general circulation in one of the principal zones of westerly winds, characterised by large wavelength (c. 6000 km), significant amplitude (c. 3000 km) and slow movement, which can be both eastward and westward relative to the Earth. In the ocean, similar waves have a wavelength of an order of a few hundred kilometres and nearly always move westward relative to the Earth.

Sea-floor spreading   The process by which lithospheric plates either side of an ocean ridge grow by the addition of new material as the plates either side of the ridge move apart.

Solar radiation   The amount of radiation or energy received from the Sun at any given point. The 'solar constant' is a measure of solar radiation (~ 1370 W m^{-2}), at a point just outside the Earth's atmosphere, located on a surface that is perpendicular to the line of radiation, and measured when the Earth is at its mean orbital distance from the Sun.

Stadial   A period during **Glacial epochs** when the ice sheets advanced to lower latitudes.

Stevenson shelter   A standard housing for ground-level meteorological instruments designed to ensure that reliable shade temperatures are measured.

Stratosphere   A region of the upper atmosphere, which extends from the tropopause to about 50 km above the Earth's surface, and where the temperature rises slowly with altitude. The properties of the stratosphere include very little vertical mixing, strong horizontal motions and low water vapour content compared to the troposphere.

Thermocline   In the ocean a region of rapidly changing temperature between the warm upper layer (the epilimnion) and the colder deeper water (the hypolimnion).

Thermohaline circulation   The deep-water circulation of the oceans driven by density contrasts due to variations in salinity and temperature.

Time series   Any series of observations of a physical variable that is sampled at set constant time intervals.

Troposphere   The lower atmosphere, from the ground to an altitude of about 8 km at the poles, about 12 km in mid-latitudes, and about 16 km in the tropics. Clouds and weather systems, as experienced by people, take place in the troposphere.

Typhoon   A name of Chinese origin (meaning 'great wind') applied to tropical cyclones which occur in the western Pacific Ocean. They are essentially the same as hurricanes in the Atlantic and cyclones in the Bay of Bengal.

Uniformatarinism   The principle that geological events can be explained by processes observable today (the present is the key to the past). This assumes these processes have not changed during geological time.

Variance   The mean of the sum of squared deviations of a set of observations from the corresponding mean.

Volcanism   The phenomena of volcanic activity. Large volcanic eruptions spew massive amounts of ash into the atmosphere that absorb solar radiation, thus potentially generating a cooling effect on planetary temperatures. At the same time, volcanoes release carbon dioxide and sulphur dioxide, and decrease stratospheric concentrations of ozone.

Weathering   The chemical or physical breakdown of rocks at the surface by atmospheric agents and physical processes.

Younger Dryas (12,900–11,600 BP)   A sudden, abrupt cold episode, which interrupted the sustained warming trend between the Last Glacial Maximum and the Holocene.

INDEX

A suffix F refers to a figure and T to a table.